The Best American Science Writing 2000

The Best American
SCIENCE WRITING
2000

GUEST EDITOR: JAMES GLEICK

Series Editor: Jesse Cohen

THE ECCO PRESS
An Imprint of HarperCollins*Publishers*

FIRST EDITION

Designed by Cassandra J. Pappas

Library of Congress Cataloging-in-Publication Data has been applied for.

ISBN 0-06-019734-X

ISBN 0-06-095736-0 (pbk.)

00   01   02   03   04   BVG   10   9   8   7   6   5   4   3   2

Contents

Introduction by James Gleick

I USED TO HAVE a pretty good idea what science writing is. The process of assembling this volume has left me in a state of uncertainty. At the risk of alarming the reader, I confess at the outset that we never arrived at anything close to a rigorous definition. Any writing about science and scientists? (Including technology? Including "nature"?) Scientists' own presentation of their research results? Their reflections on the motivation or background or history of their work? Newspaper reports of scientific discoveries? Medical memoirs? Internet haiku ("You step in the stream, / but the water has moved on. / This page is not here")?

We decided on the Big Tent approach. How could we do otherwise, recalling that a classic of science writing from the previous century begins:

> Neutrinos, they are very small.
> They have no charge and have no mass
> And do not interact at all.
> The earth is just a silly ball
> To them, through which they simply pass,
> Like dustmaids down a drafty hall
> Or photons through a sheet of glass. . . ?

(John Updike, of course; and good background for the essay by Francis Halzen, neutrino hunter, included in this volume.) Much has been said about the gap between science and the arts—science often does wear an esoteric, unfriendly mask—but that gap is not so wide as it seems. Science has woven itself into the fabric of our lives. Our best artists—novelists, poets, screenwriters, and songwriters—have noticed this and embraced it. Our newspapers, formerly treating science as alien and obscure, came around in the closing decades of the last century to a more catholic view; they created science pages and science sections, hiring science reporters who covered their beats as aggressively as their colleagues in the police shack and the campaign bus.

And no wonder. When the century began, science was a bright promise of the future, telephony and electrification bringing palpable new prosperity. By mid-century science was also a terrible shadow, created by physicists whose noble search for the heart of the atom had brought them the power to destroy an entire planet. Unstoppable—generously supported by governments—scientists began hurling stuff into space and deciphering our genetic code. (Science fiction movies, in response, bring us combined space adventures, genetic adventures, and no shortage of combined space/genetic adventures.) With mixed emotions, then, in just the last few years, we've watched the even faster, even more profound revolutions in electronics and computing, and the internetworking of the globe. There's been a lot to write about.

So now it's the year 2000 and you grab a novel at an airport bookstore, and a few minutes later, as you hurtle subsonically through what you dimly recall is our ozone-depleted, carbon-dioxide-enriched stratosphere, you read a bit of scene-setting prose about . . . well, evolution, it seems:

> Let's just set the existence-of-God issue aside for a later volume, and just stipulate that in some way, self-replicating organisms came into existence on this planet and immediately began trying to get rid of each other, either by spamming their environments with rough copies of themselves, or by more direct means which hardly need to be belabored. Most of them failed, and their genetic legacy was erased from the universe forever, but a few found a way to survive and propagate.

This is Neal Stephenson, *Cryptonomicon.* Amusing enough, and you know what he means, because we've all picked up a rudimentary understanding of how we got where we are—thanks to science writers, if not classroom teachers.

Or have we? Meanwhile, Edward Wyatt is reporting in the *New York Times* that even New York, citadel of Yankee eggheadism, has now chartered a school that will be teaching, as a form of science, that "the theory of evolution is unproven" and that "there is scientifically valid evidence that a divine being created the universe, humans and other species less than 10,000 years ago." Whoops. What is science, again? Everyone who's not a complete idiot is supposed to remember that our species is a lot older than 10,000 years and that evolution is a firm part of a broad body of knowledge, which we rely on in one way or another when we eat a tomato or let ourselves be cured by some new vaccine or watch a big-screen space/genetics thriller. But even the *New York Times* doesn't allow itself to sort out *fact* from faith, authoritatively. So there is poor Wyatt writing about "alternative theories" and "opponents of creationism."

This stuff matters. There's a lot of confusion out there. Otherwise sane, educated people find themselves not quite knowing whether millions of their friends and neighbors have been abducted and probed by almond-eyed aliens. Not a healthy state of affairs (and I don't mean the abductions). So we felt particularly well informed, not to mention entertained, by the report in that not-quite-real newspaper, *The Onion*, headlined "Revolutionary New Insoles Combine Five Forms of Pseudoscience." It says that some of the nation's top pseudoscientists "utilize the healing power of crystals to re stimulate dead foot cells with vibrational biofeedback, a process similar to that by which medicine makes people better." We feel that the author of this gem has a wonderfully firm grip on reality. We revel in our murky definition of science writing, which lets us include it here.

For better or worse, modern life demands a certain amount of sophistication about science, if we are to function properly as individuals and as members of the polity. We don't always have the knowledge or fine judgment we need—in fact, we don't ever have it—but we can engage in a lifelong process of looking for it. Atul Gawande, whose essay on mistakes by doctors opens this volume, provides a salutary example. Medical mistakes are a timely and hotly debated issue of public policy. We worry, as citizens, about whether doctors are routinely covering up their mistakes; whether our system of medical liability hurts or helps; whether the government needs to tackle the problem aggressively. Hard as it is to sort out these painful questions, it's even harder to understand just what we mean when we talk about doctors' *mistakes*. Perhaps we tend to assume that mistakes belong to *bad* doctors; that if we could just screen out the malefactors, we could approach an ideal of mistake-free medicine. Gawande explains—or, better, *shows*—that matters are more complex. "The way that things go wrong in medicine is normally unseen and, consequently, often misunderstood," he begins, with calm understatement. "Mistakes do happen. We think of them as aberrant; they are anything but." A heart-wrenching and eye-opening drama follows.

Gawande listens to the background language of his side of science: "systems problems," "continuous quality improvement," and "process reengineering." "It is the dry language of structures, not people," he notes. That is a common affliction of science writing, too, but not Gawande's.

Our writers are also travel guides, of a sort. While Gawande ushers us into the surgical theater, a field biologist, Deborah Gordon, is out in the Arizona desert near the Chiricahua Mountains, her socks rolled up over her pant cuffs to keep out the ants she is busy studying. It's no coincidence that Douglas Hofstadter, cognitive scientist and expert on artificial intelligence, wrote about ants two decades earlier in his monumental *Gödel, Escher, Bach*. He and Gordon have precisely the same analogy in mind. "The living world is structured in layers," Gordon writes in the essay reprinted here, "from molecules to cells to organs to individuals to populations to ecosystems. A fundamental question in biology is how events at those different levels are related. A thought happens when electrical impulses move around the tangle of neurons in a brain, but a thought is something more than, and something other than, neurons." Hofstadter himself, meanwhile, has been taking his work on analogy to a new level, as we learn in his breathtaking essay, "Analogy as the Core of Cognition." Maybe Gordon's writing could be called a memoir of field work. Maybe Hofstadter's has to be called, not to put too fine a point on it, *philosophy*. They're both illuminating, and somehow they're both working the same territory.

So, for that matter, is Peter Galison, nominally at a distant remove in the dry temples of academic history of science. This isn't an area normally identified with good writing, at least by our nonacademic lights. But Galison is an exception, and we had to stretch our definition yet again, to include his erudite, scholarly, heavily footnoted essay, "Einstein's Clocks: The Place of Time." He does no less than rewrite the history of special relativity here, and he does it by embracing domains left out from the standard accounts: Einstein's work at the Swiss Patent Office, often seen as an irrelevant sideshow; and the workings of clocks—*real* clocks—along railroad lines from Paris to Geneva. It was an electrochronometric world, after all, and Galison helps us see it whole. "In looking up," he writes, "—to the metaphysics of Einstein's operationalized distant simultaneity, to the shifting culture of space, time, and motion—we see down—to the wires, gears, and pulses passing through the Bern patent office."

He has poetry in his soul, it turns out, as do each of the writers whose work inspired this collection. And it's a lucky thing. We need the news they're delivering. The more we read this year, the more we saw that our technocratic age requires urgent messages from the sometimes baffling, sometimes tumultuous frontier of knowledge—the place we call science, whether or not we can define it.

The Best American Science Writing 2000

ATUL GAWANDE

When Doctors Make Mistakes

FROM *THE NEW YORKER*

A study released by the National Academy of Sciences in November 1999 reported that medical errors caused between 44,000 and 98,000 deaths a year. Congress held hearings to investigate its findings and President Clinton ordered hospitals to monitor errors and report them to a federal agency. Months before, readers of The New Yorker were introduced to the subject through the courageous reporting of a young surgical resident. Atul Gawande's bracing first-person account of life-and-death decision-making in the emergency room puts a human face on this complex and urgent issue.

I—Crash Victim

At 2 A.M. on a crisp Friday in winter, I was in sterile gloves and gown, pulling a teenage knifing victim's abdomen open, when my pager sounded. "Code Trauma, three minutes," the operating-room nurse said, reading aloud from my pager display. This meant that an ambulance would be bringing another trauma patient to the hospital momentarily, and, as the surgical resident on duty for emergencies, I would have to be present for the patient's arrival. I stepped back from the table and took off my gown. Two other surgeons were working on the knifing victim: Michael Ball, the attending (the staff surgeon in charge of the case), and David Hernandez, the chief resident (a general surgeon in his last of five years of training). Ordinarily, these two would have come later to help with the trauma, but they were stuck here. Ball, a dry, imperturbable forty-two-year-old Texan, looked over to me as I headed for the door. "If you run into any trouble, you call, and one of us will peel away," he said.

I did run into trouble. In telling this story, I have had to change significant details about what happened (including the names of the participants and aspects of my role), but I have tried to stay as close to the actual events as I could while protecting the patient, myself, and the rest of the staff. The way that things go wrong in medicine is normally unseen and, consequently, often misunderstood. Mistakes do happen. We think of them as aberrant; they are anything but.

The emergency room was one floor up, and, taking the stairs two at a time, I arrived just as the emergency medical technicians wheeled in a woman who appeared to be in her thirties and to weigh more than two hundred pounds. She lay motionless on a hard orange plastic spinal board—eyes closed, skin pale, blood running out of her nose. A nurse directed the crew into Trauma Bay 1, an examination room outfitted like an O.R., with green tiles on the wall, monitoring devices, and space for portable X-ray equipment. We lifted her onto the bed and then went to work. One nurse began cutting off the woman's clothes. Another took vital signs. A third inserted a large-bore intravenous line into her right arm. A surgical intern put a Foley catheter into her bladder. The emergency-medicine attending was Samuel Johns, a gaunt, Ichabod Crane–like man in his fifties. He was standing to one side with his arms crossed, observing, which was a sign that I could go ahead and take charge.

If you're in a hospital, most of the "moment to moment" doctoring you get is from residents—physicians receiving specialty training and a small income in exchange for their labor. Our responsibilities depend on our level of training, but we're never entirely on our own: there's always an attending, who oversees our decisions. That night, since Johns was the attending and was responsible for the patient's immediate management, I took my lead from him. But he wasn't a surgeon, and so he relied on me for surgical expertise.

"What's the story?" I asked.

An E.M.T. rattled off the details: "Unidentified white female unrestrained driver in high-speed rollover. Ejected from the car. Found unresponsive to pain. Pulse a hundred, B.P. a hundred over sixty, breathing at thirty on her own . . ."

As he spoke, I began examining her. The first step in caring for a trauma patient is always the same. It doesn't matter if a person has been shot eleven times or crushed by a truck or burned in a kitchen fire. The first thing you do is make sure that the patient can breathe without difficulty. This woman's breaths were shallow and rapid. An oximeter, by means of a sensor placed on her finger, measured the oxygen saturation of her blood. The "O_2 sat" is normally more than ninety-five percent for a patient breathing room air. The woman was wearing a face mask with oxygen turned up full blast, and her sat was only ninety percent.

"She's not oxygenating well," I announced in the flattened-out, wake-me-up-when-something-interesting-happens tone that all surgeons have acquired by about three months into residency. With my fingers, I verified that there wasn't any object in her mouth that would obstruct her airway; with a stethoscope, I confirmed that neither lung had collapsed. I got hold of a bag mask, pressed its clear facepiece over her nose and mouth, and squeezed the bellows, a kind of balloon with a one-way valve, shooting a litre of air into her with each compression. After a minute or so, her oxygen came up to a comfortable ninety-eight percent. She obviously needed our help with breathing. "Let's tube her," I said. That meant putting a tube down through her vocal cords and into her trachea, which would insure a clear airway and allow for mechanical ventilation.

Johns, the attending, wanted to do the intubation. He picked up a Mac 3 laryngoscope, a standard but fairly primitive-looking L-shaped metal instrument for prying open the mouth and throat, and slipped the shoehornlike blade deep into her mouth and down to her larynx. Then he yanked the handle up toward the ceiling to pull her tongue out of the way, open her mouth and throat, and reveal the vocal cords, which sit like fleshy tent flaps at the entrance to the trachea. The patient didn't wince or gag: she was still out cold.

"Suction!" he called. "I can't see a thing."

He sucked out about a cup of blood and clot. Then he picked up the endotracheal tube—a clear rubber pipe about the diameter of an index finger and three times as long—and tried to guide it between her cords. After a minute, her sat started to fall.

"You're down to seventy percent," a nurse announced.

Johns kept struggling with the tube, trying to push it in, but it banged vainly against the cords. The patient's lips began to turn blue.

"Sixty percent," the nurse said.

Johns pulled everything out of the patient's mouth and fitted the bag mask back on. The oximeter's luminescent-green readout hovered at sixty for a moment and then rose steadily, to ninety-seven percent. After a few minutes, he took the mask off and again tried to get the tube in. There was more blood, and there may have been some swelling, too: all the poking down the throat was probably not helping. The sat fell to sixty percent. He pulled out and bagged her until she returned to ninety-five percent.

When you're having trouble getting the tube in, the next step is to get specialized expertise. "Let's call anesthesia," I said, and Johns agreed. In the meantime, I continued to follow the standard trauma protocol: completing the examination and ordering fluids, lab tests, and X-rays. Maybe five minutes passed as I worked.

The patient's sats drifted down to ninety-two percent—not a dramatic change but definitely not normal for a patient who is being manually ventilated. I checked to see if the sensor had slipped off her finger. It hadn't. "Is the oxygen up full blast?" I asked a nurse.

"It's up all the way," she said.

I listened again to the patient's lungs—no collapse. "We've got to get her tubed," Johns said. He took off the oxygen mask and tried again.

Somewhere in my mind, I must have been aware of the possibility that her airway was shutting down because of vocal-cord swelling or blood. If it was, and we were unable to get a tube in, then the only chance she'd have to survive would be an emergency tracheostomy: cutting a hole in her neck and inserting a breathing tube into her trachea. Another attempt to intubate her might even trigger a spasm of the cords and a sudden closure of the airway—which is exactly what did happen.

If I had actually thought this far along, I would have recognized how ill-prepared I was to do an emergency "trache." Of the people in the room, it's true, I had the most experience doing tracheostomies, but that wasn't saying much. I had been the assistant surgeon in only about half a dozen, and all but one of them had been non-emergency cases, employing techniques that were not designed for speed. The exception was a practice emergency trache I had done on a goat. I should have immediately called Dr. Ball for backup. I should have got the trache equipment out—lighting, suction, sterile instruments—just in case. Instead of hurrying the effort to get the patient intubated because of a mild drop in saturation, I should have asked Johns to wait until I had help nearby. I might even have recognized that she was already losing her airway. Then I could have grabbed a knife and started cutting her a tracheostomy while things were still relatively stable and I had time to proceed slowly. But for whatever reasons—hubris, inattention, wishful thinking, hesitation, or the uncertainty of the moment—I let the opportunity pass.

Johns hunched over the patient, intently trying to insert the tube through her vocal cords. When her sat once again dropped into the sixties, he stopped and put the mask back on. We stared at the monitor. The numbers weren't coming up. Her lips were still blue. Johns squeezed the bellows harder to blow more oxygen in.

"I'm getting resistance," he said.

The realization crept over me: this was a disaster. "Damn it, we've lost her airway," I said. "Trache kit! Light! Somebody call down to O.R. 25 and get Ball up here!"

People were suddenly scurrying everywhere. I tried to proceed deliberately, and not let panic take hold. I told the surgical intern to get a sterile gown and gloves on. I took a bactericidal solution off a shelf and dumped a whole bottle of yellow-brown liquid on the patient's neck. A nurse unwrapped the tracheostomy kit—a sterilized set of drapes and instruments. I pulled on a gown and a new pair of gloves while trying to think through the steps. This is simple, really, I tried to tell myself. At the base of the thyroid cartilage, the Adam's apple, is a little gap in which you find a thin, fibrous covering called the cricothyroid membrane. Cut through that and—voilà! You're in the trachea. You slip through the hole a four-inch plastic tube shaped like a plumber's elbow joint, hook it up to oxygen and a ventilator, and she's all set. Anyway, that was the theory.

I threw some drapes over her body, leaving the neck exposed. It looked as thick as a tree. I felt for the bony prominence of the thyroid cartilage. But I couldn't feel anything through the rolls of fat. I was beset by uncertainty—where should I cut? should I make a horizontal or a vertical incision?—and I hated myself for it. Surgeons never dithered, and I was dithering.

"I need better light," I said.

Someone was sent out to look for one.

"Did anyone get Ball?" I asked. It wasn't exactly an inspiring question.

"He's on his way," a nurse said.

There wasn't time to wait. Four minutes without oxygen would lead to permanent brain damage, if not death. Finally, I took the scalpel and cut. I just cut. I made a three-inch left-to-right swipe across the middle of the neck, following the procedure I'd learned for elective cases. I figured that if I worked through the fat I might be able to find the membrane in the wound. Dissecting down with scissors while the intern held the wound open with retractors, I hit a vein. It didn't let loose a lot of blood, but there was enough to fill the wound: I couldn't see anything. The intern put a finger on the bleeder. I called for suction. But the suction wasn't working; the tube was clogged with the clot from the intubation efforts.

"Somebody get some new tubing," I said. "And where's the light?"

Finally, an orderly wheeled in a tall overhead light, plugged it in, and flipped on the switch. It was still too dim; I could have done better with a flashlight.

I wiped up the blood with gauze, then felt around in the wound with my fingertips. This time, I thought I could feel the hard ridges of the thyroid cartilage and, below it, the slight gap of the cricothyroid membrane, though I couldn't be sure. I held my place with my left hand.

James O'Connor, a silver-haired, seen-it-all anesthesiologist, came into the room. Johns gave him a quick rundown on the patient and let him take over bagging her.

Holding the scalpel in my right hand like a pen, I stuck the blade down into the wound at the spot where I thought the thyroid cartilage was. With small, sharp strokes—working blindly, because of the blood and the poor light—I cut down through the overlying fat and tissue until I felt the blade scrape against the almost bony cartilage. I searched with the tip of the knife, walking it along until I felt it reach a gap. I hoped it was the cricothyroid membrane, and pressed down firmly. Then I felt the tissue suddenly give, and I cut an inch-long opening.

When I put my index finger into it, it felt as if I were prying open the jaws of a stiff clothespin. Inside, I thought I felt open space. But where were the sounds of moving air that I expected? Was this deep enough? Was I even in the right place?

"I think I'm in," I said, to reassure myself as much as anyone else.

"I hope so," O'Connor said. "She doesn't have much longer."

I took the tracheostomy tube and tried to fit it in, but something seemed to be blocking it. I twisted it and turned it, and finally jammed it in. Just then, Ball, the surgical attending, arrived. He rushed up to the bed and leaned over for a look. "Did you get it?" he asked. I said that I thought so. The bag mask was plugged onto the open end of the trache tube. But when the bellows were compressed the air just gurgled out of the wound. Ball quickly put on gloves and a gown.

"How long has she been without an airway?" he asked.

"I don't know. Three minutes."

Ball's face hardened as he registered that he had about a minute in which to turn things around. He took my place and summarily pulled out the trache tube. "God, what a mess," he said. "I can't see a thing in this wound. I don't even know if you're in the right place. Can we get better light and suction?" New suction tubing was found and handed to him. He quickly cleaned up the wound and went to work.

The patient's sat had dropped so low that the oximeter couldn't detect it anymore. Her heart rate began slowing down—first to the sixties and then to the forties. Then she lost her pulse entirely. I put my hands together on her chest, locked my elbows, leaned over her, and started doing chest compressions.

Ball looked up from the patient and turned to O'Connor. "I'm not going to get her an airway in time," he said. "You're going to have to try from above." Essentially, he was admitting my failure. Trying an oral intubation again was pointless—just something to do instead of watching her die. I was stricken, and concentrated on doing chest compressions, not looking at anyone. It was over, I thought.

And then, amazingly, O'Connor: "I'm in." He had managed to slip a pediatric-size endotracheal tube through the vocal cords. In thirty seconds, with oxygen being manually ventilated through the tube, her heart was back, racing at a hundred and twenty beats a minute. Her sat registered at sixty and then climbed. Another thirty seconds and it was at ninety-seven percent. All the people in the room exhaled, as if they, too, had been denied their breath. Ball and I said little except to confer about the next steps for her. Then he went back downstairs to finish working on the stab-wound patient still in the O.R.

We eventually identified the woman, whom I'll call Louise Williams; she was thirty-four years old and lived alone in a nearby suburb. Her alcohol level on arrival had been three times the legal limit, and had probably contributed to her unconsciousness. She had a concussion, several lacerations, and significant soft-tissue damage. But X-rays and scans revealed no other injuries from the crash. That night, Ball and Hernandez brought her to the O.R. to fit her with a proper tracheostomy. When Ball came out and talked to family members, he told them of the dire condition she was in when she arrived, the difficulties "we" had had getting access to her airway, the disturbingly long period of time that she had gone without oxygen, and thus his uncertainty about how much brain function she still possessed. They listened without protest; there was nothing for them to do but wait.

II—The Banality of Error

TO MUCH OF THE PUBLIC—and certainly to lawyers and the media—medical error is a problem of bad physicians. Consider some other surgical mishaps. In one, a general surgeon left a large metal instrument in a patient's abdomen, where it tore through the bowel and the wall of the bladder. In another, a cancer surgeon biopsied the wrong part of a woman's breast and thereby delayed her diagnosis of cancer for months. A cardiac surgeon skipped a small but key step during a heart-valve operation, thereby killing the patient. A surgeon saw a man racked with abdominal pain in the emergency room and, without taking a C.T. scan, assumed that the man had a kidney stone; eighteen hours later, a scan showed a rupturing abdominal aortic aneurysm, and the patient died not long afterward.

How could anyone who makes a mistake of that magnitude be allowed to practice medicine? We call such doctors "incompetent," "unethical," and "negligent." We want to see them punished. And so we've wound up with the public system we have for dealing with error: malpractice lawsuits, media scandal, suspensions, firings.

There is, however, a central truth in medicine that complicates this tidy vision of misdeeds and misdoers: *All* doctors make terrible mistakes. Consider the cases I've just described. I gathered them simply by asking respected surgeons I know—surgeons at top medical schools—to tell me about mistakes they had made just in the past year. Every one of them had a story to tell.

In 1991, *The New England Journal of Medicine* published a series of landmark papers from a project known as the Harvard Medical Practice Study—a review of more than thirty thousand hospital admissions in New York State. The study found that nearly four percent of hospital patients suffered complications from treatment which prolonged their hospital stay or resulted in disability or death, and that two-thirds of such complications were due to errors in care. One in four, or one percent of admissions, involved actual negligence. It was estimated that, nationwide, a hundred and twenty thousand patients die each year at least partly as a result of errors in care. And subsequent investigations around the country have confirmed the ubiquity of error. In one small study of how clinicians perform when patients have a sudden cardiac arrest, twenty-seven of thirty clinicians made an error in using the defibrillator; they may have charged it incorrectly or lost valuable time trying to figure out how to work a particular model. According to a 1995 study, mistakes in administering drugs—giving the wrong drug or the wrong dose, say—occur, on the average, about once for every hospital admission, mostly without ill effects, but one percent of the time with serious consequences.

If error were due to a subset of dangerous doctors, you might expect malpractice cases to be concentrated among a small group, but in fact they follow a uniform, bell-shaped distribution. Most surgeons are sued at least once in the course of their careers. Studies of specific types of error, too, have found that repeat offenders are not the problem. The fact is that virtually everyone who cares for hospital patients will make serious mistakes, and even commit acts of negligence, every year. For this reason, doctors are seldom outraged when the press reports yet another medical horror story. They usually have a different reaction: *That could be me.* The important question isn't how to keep bad physicians from harming patients; it's how to keep good physicians from harming patients.

Medical-malpractice suits are a remarkably ineffective remedy. Troyen Brennan, a Harvard professor of law and public health, points out that research has consistently failed to find evidence that litigation reduces medical-error rates. In part, this may be because the weapon is so imprecise. Brennan led several studies following up on the patients in the Harvard Medical Practice Study. He found that fewer than two percent of the patients who had received substandard care ever filed suit. Conversely, only a small minority among the patients who did sue had in fact been the victims of negligent care. And a patient's likelihood of winning a suit depended primarily on how poor his or her outcome was, regardless of whether that outcome was caused by disease or unavoidable risks of care.

The deeper problem with medical-malpractice suits, however, is that by demonizing errors they prevent doctors from acknowledging and discussing them publicly. The tort system makes adversaries of patient and physician, and pushes each to offer a heavily slanted version of events. When things go wrong, it's almost impossible for a physician to talk to a patient honestly about mistakes. Hospital lawyers warn doctors that, although they must, of course, tell patients about complications that occur, they are never to intimate that they were at fault, lest the "confession" wind up in court as damning evidence in a black-and-white morality tale. At most, a doctor might say, "I'm sorry that things didn't go as well as we had hoped."

There is one place, however, where doctors can talk candidly about their mistakes, if not with patients, then at least with one another. It is called the Morbidity and Mortality Conference—or, more simply, M. & M.—and it takes place, usually once a week, at nearly every academic hospital in the country. This institution survives because laws protecting its proceedings from legal discovery have stayed on the books in most states, despite frequent challenges. Surgeons, in particular, take the M. & M. seriously. Here they can gather behind closed doors to review the mistakes, complications, and deaths that occurred on their watch, determine responsibility, and figure out what to do differently next time.

III—Show and Tell

AT MY HOSPITAL, we convene every Tuesday at five o'clock in a steep, plush amphitheatre lined with oil portraits of the great doctors whose achievements we're meant to live up to. All surgeons are expected to attend, from the interns to the chairman of surgery; we're also joined by medical students doing their surgery "rotation." An M. & M. can include almost a hundred people. We file in, pick up a photocopied list of cases to be discussed, and take our seats. The front row is occupied by the most senior surgeons: terse, serious men, now out of their scrubs and in dark suits, lined up like a panel of senators at a hearing. The chairman is a leonine presence in the seat closest to the plain wooden podium from which each case is presented. In the next few rows are the remaining surgical attendings; these tend to be younger, and several of them are women. The chief residents have put on long white coats and usually sit in the side rows. I join the mass of other residents, all of us in short white coats and green scrub pants, occupying the back rows.

For each case, the chief resident from the relevant service—cardiac, vascular, trauma, and so on—gathers the information, takes the podium, and tells the story. Here's a partial list of cases from a typical week (with a few changes to protect confidentiality): a sixty-eight-year-old man who bled to death after heart-valve surgery; a forty-seven-year-old woman who had to have a reoperation because of infection following an arterial bypass done in her left leg; a forty-four-year-old woman who had to have bile drained from her abdomen after gall-bladder surgery; three patients who had to have reoperations for bleeding following surgery; a sixty-three-year-old man who had a cardiac arrest following heart-bypass surgery; a sixty-six-year-old woman whose sutures suddenly gave way in an abdominal wound and nearly allowed her intestines to spill out. Ms. Williams's case, my failed tracheostomy, was just one case on a list like this. David Hernandez, the chief trauma resident, had subsequently reviewed the records and spoken to me and others involved. When the time came, it was he who stood up front and described what had happened.

Hernandez is a tall, rollicking, good old boy who can tell a yarn, but M. & M. presentations are bloodless and compact. He said something like: "This was a thirty-four-year-old female unrestrained driver in a high-speed rollover. The patient apparently had stable vitals at the scene but was unresponsive, and brought in by ambulance unintubated. She was G.C.S. 7 on arrival." G.C.S. stands for the Glasgow Coma Scale, which rates the severity of head injuries, from three to fifteen. G.C.S. 7 is in the comatose range. "Attempts to intubate were made without success in the E.R. and may have contributed to airway closure. A cricothyroidotomy was attempted without success."

These presentations can be awkward. The chief residents, not the attendings, determine which cases to report. That keeps the attendings honest—no one can cover up mistakes—but it puts the chief residents, who are, after all, underlings, in a delicate position. The successful M. & M. presentation inevitably involves a certain elision of detail and a lot of passive verbs. No one screws up a cricothyroidotomy. Instead, "a cricothyroidotomy was attempted without success." The message, however, was not lost on anyone.

Hernandez continued, "The patient arrested and required cardiac compressions. Anesthesia was then able to place a pediatric E.T. tube and the patient recovered stable vitals. The tracheostomy was then completed in the O.R."

So Louise Williams had been deprived of oxygen long enough to go into cardiac arrest, and everyone knew that meant she could easily have suffered a disabling stroke or been left a vegetable. Hernandez concluded with the fortunate aftermath: "Her workup was negative for permanent cerebral damage or other major injuries. The tracheostomy was removed on Day 2. She was discharged to home in good condition on Day 3." To the family's great relief, and mine, she had woken up in the morning a bit woozy but hungry, alert, and mentally intact. In a few weeks, the episode would heal to a scar.

But not before someone was called to account. A front-row voice immediately thundered, "What do you mean, 'A cricothyroidotomy was attempted without success?' " I sank into my seat, my face hot.

"This was my case," Dr. Ball volunteered from the front row. It is how every attending begins, and that little phrase contains a world of surgical culture. For all the talk in business schools and in corporate America about the virtues of "flat organizations," surgeons maintain an old-fashioned sense of hierarchy. When things go wrong, the attending is expected to take full responsibility. It makes no difference whether it was the resident's hand that slipped and lacerated an aorta; it doesn't matter whether the attending was at home in bed when a nurse gave a wrong dose of medication. At the M. & M., the burden of responsibility falls on the attending.

Ball went on to describe the emergency attending's failure to intubate Williams and his own failure to be at her bedside when things got out of control. He described the bad lighting and her extremely thick neck, and was careful to make those sound not like excuses but merely like complicating factors. Some attendings shook their heads in sympathy. A couple of them asked questions to clarify certain details. Throughout, Ball's tone was objective, detached. He had the air of a CNN newscaster describing unrest in Kuala Lumpur.

As always, the chairman, responsible for the over-all quality of our surgery service, asked the final question. What, he wanted to know, would Ball have done differently? Well, Ball replied, it didn't take long to get the stab-wound patient under control in the O.R., so he probably should have sent Hernandez up to the E.R. at that point or let Hernandez close the abdomen while he himself came up. People nodded. Lesson learned. Next case.

At no point during the M. & M. did anyone question why I had not called for help sooner or why I had not had the skill and knowledge that Williams needed. This is not to say that my actions were seen as acceptable. Rather, in the hierarchy, addressing my errors was Ball's role. The day after the disaster, Ball had caught me in the hall and taken me aside. His voice was more wounded than angry as he went through my specific failures. First, he explained, in an emergency tracheostomy it might have been better to do a vertical neck incision; that would have kept me out of the blood vessels, which run up and down—something I should have known at least from my reading. I might have had a much easier time getting her an airway then, he said. Second, and worse to him than mere ignorance, he didn't understand why I hadn't called him when there were clear signs of airway trouble developing. I offered no excuses. I promised to be better prepared for such cases and to be quicker to ask for help.

Even after Ball had gone down the fluorescent-lit hallway, I felt a sense of shame like a burning ulcer. This was not guilt: guilt is what you feel when you have done something wrong. What I felt was shame: *I* was what was wrong. And yet I also knew that a surgeon can take such feelings too far. It is one thing to be aware of one's limitations. It is another to be plagued by self-doubt. One surgeon with a national reputation told me about an abdominal operation in which he had lost control of bleeding while he was removing what turned out to be a benign tumor and the patient had died. "It was a clean kill," he said. Afterward, he could barely bring himself to operate. When he did operate, he became tentative and indecisive. The case affected his performance for months.

Even worse than losing self-confidence, though, is reacting defensively. There are surgeons who will see faults everywhere except in themselves. They have no questions and no fears about their abilities. As a result, they learn nothing from their mistakes and know nothing of their limitations. As one surgeon told me, it is a rare but alarming thing to meet a surgeon without fear. "If you're not a little afraid when you operate," he said, "you're bound to do a patient a grave disservice."

The atmosphere at the M. & M. is meant to discourage both attitudes—self-doubt and denial—for the M. & M. is a cultural ritual that inculcates in surgeons a "correct" view of mistakes. "What would you do differently?" a chairman asks concerning cases of avoidable complications. "Nothing" is seldom an acceptable answer.

In its way, the M. & M. is an impressively sophisticated and human institution. Unlike the courts or the media, it recognizes that human error is generally not something that can be deterred by punishment. The M. & M. sees avoiding error as largely a matter of will—of staying sufficiently informed and alert to anticipate the myriad ways that things can go wrong and then trying to head off each potential problem before it happens. Why do things go wrong? Because, doctors say, making them go right is hard stuff. It isn't damnable that an error occurs, but there is some shame to it. In fact, the M. & M.'s ethos can seem paradoxical. On the one hand, it reinforces the very American idea that error is intolerable. On the other hand, the very existence of the M. & M., its place on the weekly schedule, amounts to an acknowledgment that mistakes are an inevitable part of medicine.

BUT WHY DO THEY HAPPEN so often? Lucian Leape, medicine's leading expert on error, points out that many other industries—whether the task is manufacturing semiconductors or serving customers at the Ritz-Carlton—simply wouldn't countenance error rates like those in hospitals. The aviation industry has reduced the frequency of operational errors to one in a hundred thousand flights, and most of those errors have no harmful consequences. The buzzword at General Electric these days is "Six Sigma," meaning that its goal is to make product defects so rare that in statistical terms they are more than six standard deviations away from being a matter of chance—almost a one-in-a-million occurrence.

Of course, patients are far more complicated and idiosyncratic than airplanes, and medicine isn't a matter of delivering a fixed product or even a catalogue of products; it may well be more complex than just about any other field of human endeavor. Yet everything we've learned in the past two decades—from cognitive psychology, from "human factors" engineering, from studies of disasters like Three Mile Island and Bhopal—has yielded the same insights: not only do all human beings err but they err frequently and in predictable, patterned ways. And systems that do not adjust for these realities can end up exacerbating rather than eliminating error.

The British psychologist James Reason argues, in his book *Human Error,* that our propensity for certain types of error is the price we pay for the brain's remarkable ability to think and act intuitively—to sift quickly through the sensory information that constantly bombards us without wasting time trying to work through every situation anew. Thus systems that rely on human perfection present what Reason calls "latent errors"—errors waiting to happen. Medicine teems with examples. Take writing out a prescription, a rote procedure that relies on memory and attention, which we know are unreliable. Inevitably, a physician will sometimes specify the wrong dose or the wrong drug. Even when the prescription is written correctly, there's a risk that it will be misread. (Computerized ordering systems can almost eliminate errors of this kind, but only a small minority of hospitals have adopted them.) Medical equipment, which manufacturers often build without human operators in mind, is another area rife with latent errors: one reason physicians are bound to have problems when they use cardiac defibrillators is that the devices have no standard design. You can also make the case that onerous workloads, chaotic environments, and inadequate team communication all represent latent errors in the system.

James Reason makes another important observation: disasters do not simply occur; they evolve. In complex systems, a single failure rarely leads to harm. Human beings are impressively good at adjusting when an error becomes apparent, and systems often have built-in defenses. For example, pharmacists and nurses routinely check and counter-check physicians' orders. But errors do not always become apparent, and backup systems themselves often fail as a result of latent errors. A pharmacist forgets to check one of a thousand prescriptions. A machine's alarm bell malfunctions. The one attending trauma surgeon available gets stuck in the operating room. When things go wrong, it is usually because a series of failures conspire to produce disaster.

The M. & M. takes none of this into account. For that reason, many experts see it as a rather shabby approach to analyzing error and improving performance in medicine. It isn't enough to ask what a clinician could or should have done differently so that he and others may learn for next time. The doctor is often only the final actor in a chain of events that set him or her up to fail. Error experts, therefore, believe that it's the process, not the individuals in it, which requires closer examination and correction. In a sense, they want to industrialize medicine. And they can already claim one success story: the specialty of anesthesiology, which has adopted their precepts and seen extraordinary results.

IV—Nearly Perfect

AT THE CENTER of the emblem of the American Society of Anesthesiologists is a single word: "Vigilance." When you put a patient to sleep under general anesthesia, you assume almost complete control of the patient's body. The body is paralyzed, the brain rendered unconscious, and machines are hooked up to control breathing, heart rate, blood pressure—all the vital functions. Given the complexity of the machinery and of the human body, there are a seemingly infinite number of ways in which things can go wrong, even in minor surgery. And yet anesthesiologists have found that if problems are detected they can usually be solved. In the nineteen-forties, there was only one death resulting from anesthesia in every twenty-five hundred operations, and between the nineteen-sixties and the nineteen-eighties the rate had stabilized at one or two in every ten thousand operations.

But Ellison (Jeep) Pierce had always regarded even that rate as unconscionable. From the time he began practicing, in 1960, as a young anesthesiologist out of North Carolina and the University of Pennsylvania, he had maintained a case file of details from all the deadly anesthetic accidents he had come across or participated in. But it was one case in particular that galvanized him. Friends of his had taken their eighteen-year-old daughter to the hospital to have her wisdom teeth pulled, under general anesthesia. The anesthesiologist inserted the breathing tube into her esophagus instead of her trachea, which is a relatively common mishap, and then failed to spot the error, which is not. Deprived of oxygen, she died within minutes. Pierce knew that a one-in-ten-thousand death rate, given that anesthesia was administered in the United States an estimated thirty-five million times each year, meant thirty-five hundred avoidable deaths like that one.

In 1982, Pierce was elected vice-president of the American Society of Anesthesiologists and got an opportunity to do something about the death rate. The same year, ABC's "20/20" aired an exposé that caused a considerable stir in his profession. The segment began, "If you are going to go into anesthesia, you are going on a long trip, and you should not do it if you can avoid it in any way. General anesthesia [is] safe most of the time, but there are dangers from human error, carelessness, and a critical shortage of anesthesiologists. This year, six thousand patients will die or suffer brain damage." The program presented several terrifying cases from around the country. Between the small crisis that the show created and the sharp increases in physicians' malpractice-insurance premiums at that time, Pierce was able to mobilize the Society of Anesthesiologists around the problem of error.

He turned for ideas not to a physician but to an engineer named Jeffrey Cooper, the lead author of a ground-breaking 1978 paper entitled "Preventable Anesthesia Mishaps: A Study of Human Factors." An unassuming, fastidious man, Cooper had been hired in 1972, when he was twenty-six years old, by the Massachusetts General Hospital bioengineering unit, to work on developing machines for anesthesiology researchers. He gravitated toward the operating room, however, and spent hours there observing the anesthesiologists, and one of the first things he noticed was how poorly the anesthesia machines were designed. For example, a clockwise turn of a dial decreased the concentration of potent anesthetics in about half the machines but increased the concentration in the other half. He decided to borrow a technique called "critical incident analysis"—which had been used since the nineteen-fifties to analyze mishaps in aviation—in an effort to learn how equipment might be contributing to errors in anesthesia. The technique is built around carefully conducted interviews, designed to capture as much detail as possible about dangerous incidents: how specific accidents evolved and what factors contributed to them. This information is then used to look for patterns among different cases.

Getting open, honest reporting is crucial. The Federal Aviation Administration has a formalized system for analyzing and reporting dangerous aviation incidents, and its enormous success in improving airline safety rests on two cornerstones. Pilots who report an incident within ten days have automatic immunity from punishment, and the reports go to a neutral, outside agency, NASA, which has no interest in using the information against individual pilots. For Jeffrey Cooper, it was probably an advantage that he was an engineer, and not a physician, so that anesthesiologists regarded him as a discreet, unthreatening interviewer.

The result was the first in-depth, scientific look at errors in medicine. His detailed analysis of three hundred and fifty-nine errors provided a view of the profession unlike anything that had been seen before. Contrary to the prevailing assumption that the start of anesthesia ("takeoff") was the most dangerous part, anesthesiologists learned that incidents tended to occur in the middle of anesthesia, when vigilance waned. The most common kind of incident involved errors in maintaining the patient's breathing, and these were usually the result of an undetected disconnection or misconnection of the breathing tubing, mistakes in managing the airway, or mistakes in using the anesthesia machine. Just as important, Cooper enumerated a list of contributory factors, including inadequate experience, inadequate familiarity with equipment, poor communication among team members, haste, inattention, and fatigue.

The study provoked widespread debate among anesthesiologists, but there was no concerted effort to solve the problems until Jeep Pierce came along. Through the anesthesiology society at first, and then through a foundation that he started, Pierce directed funding into research on how to reduce the problems Cooper had identified, sponsored an international conference to gather ideas from around the world, and brought anesthesia-machine designers into safety discussions.

It all worked. Hours for anesthesiology residents were shortened. Manufacturers began redesigning their machines with fallible human beings in mind. Dials were standardized to turn in a uniform direction; locks were put in to prevent accidental administration of more than one anesthetic gas; controls were changed so that oxygen delivery could not be turned down to zero.

Where errors could not be eliminated directly, anesthesiologists began looking for reliable means of detecting them earlier. For example, because the trachea and the esophagus are so close together, it is almost inevitable that an anesthesiologist will sometimes put the breathing tube down the wrong pipe. Anesthesiologists had always checked for this by listening with a stethoscope for breath sounds over both lungs. But Cooper had turned up a surprising number of mishaps—like the one that befell the daughter of Pierce's friends—involving undetected esophageal intubations. Something more effective was needed. In fact, monitors that could detect this kind of error had been available for years, but, in part because of their expense, relatively few anesthesiologists used them. One type of monitor could verify that the tube was in the trachea by detecting carbon dioxide being exhaled from the lungs. Another type, the pulse oximeter, tracked blood-oxygen levels, thereby providing an early warning that something was wrong with the patient's breathing system. Prodded by Pierce and others, the anesthesiology society made the use of both types of monitor for every patient receiving general anesthesia an official standard. Today, anesthesia deaths from misconnecting the breathing system or intubating the esophagus rather than the trachea are virtually unknown. In a decade, the over-all death rate dropped to just one in more than two hundred thousand cases—less than a twentieth of what it had been.

And the reformers have not stopped there. David Gaba, a professor of anesthesiology at Stanford, has focussed on improving human performance. In aviation, he points out, pilot experience is recognized to be invaluable but insufficient: pilots seldom have direct experience with serious plane malfunction anymore. They are therefore required to undergo yearly training in crisis simulators. Why not doctors, too?

Gaba, a physician with training in engineering, led in the design of an anesthesia-simulation system known as the Eagle Patient Simulator. It is a life-size, computer-driven mannequin that is capable of amazingly realistic behavior. It has a circulation, a heartbeat, and lungs that take in oxygen and expire carbon dioxide. If you inject drugs into it or administer inhaled anesthetics, it will detect the type and amount, and its heart rate, its blood pressure, and its oxygen levels will respond appropriately. The "patient" can be made to develop airway swelling, bleeding, and heart disturbances. The mannequin is laid on an operating table in a simulation room equipped exactly like the real thing. Here both residents and experienced attending physicians learn to perform effectively in all kinds of dangerous, and sometimes freak, scenarios: an anesthesia-machine malfunction, a power outage, a patient who goes into cardiac arrest during surgery, and even a cesarean-section patient whose airway shuts down and who requires an emergency tracheostomy.

Though anesthesiology has unquestionably taken the lead in analyzing and trying to remedy "systems" failures, there are signs of change in other quarters. The American Medical Association, for example, set up its National Patient Safety Foundation in 1997 and asked Cooper and Pierce to serve on the board of directors. The foundation is funding research, sponsoring conferences, and attempting to develop new standards for hospital drug-ordering systems that could substantially reduce medication mistakes—the single most common type of medical error.

Even in surgery there have been some encouraging developments. For instance, operating on the wrong knee or foot or other body part of a patient has been a recurrent, if rare, mistake. A typical response has been to fire the surgeon. Recently, however, hospitals and surgeons have begun to recognize that the body's bilateral symmetry makes these errors predictable. Last year, the American Academy of Orthopedic Surgeons endorsed a simple way of preventing them: make it standard practice for surgeons to initial, with a marker, the body part to be cut before the patient comes to surgery.

The Northern New England Cardiovascular Disease Study Group, based at Dartmouth, is another success story. Though the group doesn't conduct the sort of in-depth investigation of mishaps that Jeffrey Cooper pioneered, it has shown what can be done simply through statistical monitoring. Six hospitals belong to this consortium, which tracks deaths and complications (such as wound infections, uncontrolled bleeding, and stroke) arising from heart surgery and tries to identify various risk factors. Its researchers found, for example, that there were relatively high death rates among patients who developed anemia after bypass surgery, and that anemia developed most often in small patients. The fluid used to "prime" the heart-lung machine caused the anemia, because it diluted a patient's blood, so the smaller the patient (and his or her blood supply) the greater the effect. Members of the consortium now have several promising solutions to the problem. Another study found that a group at one hospital had made mistakes in "handoffs"—say, in passing preoperative lab results to the people in the operating room. The study group solved the problem by developing a pilot's checklist for all patients coming to the O.R. These efforts have introduced a greater degree of standardization, and so reduced the death rate in those six hospitals from four percent to three percent between 1991 and 1996. That meant two hundred and ninety-three fewer deaths. But the Northern New England cardiac group, even with its narrow focus and techniques, remains an exception; hard information about how things go wrong is still scarce. There is a hodgepodge of evidence that latent errors and systemic factors may contribute to surgical errors: the lack of standardized protocols, the surgeon's inexperience, the hospital's inexperience, inadequately designed technology and techniques, thin staffing, poor teamwork, time of day, the effects of managed care and corporate medicine, and so on and so on. But which are the major risk factors? We still don't know. Surgery, like most of medicine, awaits its Jeff Cooper.

V—Getting It Right

IT WAS A ROUTINE gallbladder operation, on a routine day: on the operating table was a mother in her forties, her body covered by blue paper drapes except for her round, antiseptic-coated belly. The gallbladder is a floppy, finger-length sac of bile like a deflated olive-green balloon tucked under the liver, and when gallstones form, as this patient had learned, they can cause excruciating bouts of pain. Once we removed her gallbladder, the pain would stop.

There are risks to this surgery, but they used to be much greater. Just a decade ago, surgeons had to make a six-inch abdominal incision that left patients in the hospital for the better part of a week just recovering from the wound. Today, we've learned to take out gallbladders with a minute camera and instruments that we manipulate through tiny incisions. The operation, often done as day surgery, is known as laparoscopic cholecystectomy, or "lap chole." Half a million Americans a year now have their gallbladders removed this way; at my hospital alone, we do several hundred lap choles annually.

When the attending gave me the go-ahead, I cut a discreet inch-long semi-circle in the wink of skin just above the belly button. I dissected through fat and fascia until I was inside the abdomen, and dropped into place a "port," a half-inch-wide sheath for slipping instruments in and out. We hooked gas tubing up to a side vent on the port, and carbon dioxide poured in, inflating the abdomen until it was distended like a tire. I inserted the miniature camera. On a video monitor a few feet away, the woman's intestines blinked into view. With the abdomen inflated, I had room to move the camera, and I swung it around to look at the liver. The gallbladder could be seen poking out from under the edge.

We put in three more ports through even tinier incisions, spaced apart to complete the four corners of a square. Through the ports on his side, the attending put in two long "graspers," like small-scale versions of the device that a department-store clerk might use to get a hat off the top shelf. Watching the screen as he maneuvered them, he reached under the edge of the liver, clamped onto the gallbladder, and pulled it up into view. We were set to proceed.

Removing the gallbladder is fairly straightforward. You sever it from its stalk and from its blood supply, and pull the rubbery sac out of the abdomen through the incision near the belly button. You let the carbon dioxide out of the belly, pull out the ports, put a few stitches in the tiny incisions, slap some Band-Aids on top, and you're done. There's one looming danger, though: the stalk of the gallbladder is a branch off the liver's only conduit for sending bile to the intestines for the digestion of fats. And if you accidentally injure this main bile duct, the bile backs up and starts to destroy the liver. Between ten and twenty percent of the patients to whom this happens will die. Those who survive often have permanent liver damage and can go on to require liver transplantation. According to a standard textbook, "injuries to the main bile duct are nearly always the result of misadventure during operation and are therefore a serious reproach to the surgical profession." It is a true surgical error, and, like any surgical team doing a lap chole, we were intent on avoiding this mistake.

Using a dissecting instrument, I carefully stripped off the fibrous white tissue and yellow fat overlying and concealing the base of the gallbladder. Now we could see its broad neck and the short stretch where it narrowed down to a duct—a tube no thicker than a strand of spaghetti peeking out from the surrounding tissue, but magnified on the screen to the size of major plumbing. Then, just to be absolutely sure we were looking at the gallbladder duct and not the main bile duct, I stripped away some more of the surrounding tissue. The attending and I stopped at this point, as we always do, and discussed the anatomy. The neck of the gallbladder led straight into the tube we were eying. So it had to be the right duct. We had exposed a good length of it without a sign of the main bile duct. Everything looked perfect, we agreed. "Go for it," the attending said.

I slipped in the clip applier, an instrument that squeezes V-shaped metal clips onto whatever you put in its jaws. I got the jaws around the duct and was about to fire when my eye caught, on the screen, a little globule of fat lying on top of the duct. That wasn't necessarily anything unusual, but somehow it didn't look right. With the tip of the clip applier, I tried to flick it aside, but, instead of a little globule, a whole layer of thin unseen tissue came up, and, underneath, we saw that the duct had a fork in it. My stomach dropped. If not for that little extra fastidiousness, I would have clipped off the main bile duct.

Here was the paradox of error in medicine. With meticulous technique and assiduous effort to insure that they have correctly identified the anatomy, surgeons need never cut the main bile duct. It is a paradigm of an avoidable error. At the same time, studies show that even highly experienced surgeons inflict this terrible injury about once in every two hundred lap choles. To put it another way, I may have averted disaster this time, but a statistician would say that, no matter how hard I tried, I was almost certain to make this error at least once in the course of my career.

But the story doesn't have to end here, as the cognitive psychologists and industrial-error experts have demonstrated. Given the results they've achieved in anesthesiology, it's clear that we can make dramatic improvements by going after the process, not the people. But there are distinct limitations to the industrial cure, however necessary its emphasis on systems and structures. It would be deadly for us, the individual actors, to give up our belief in human perfectibility. The statistics may say that someday I will sever someone's main bile duct, but each time I go into a gallbladder operation I believe that with enough will and effort I can beat the odds. This isn't just professional vanity. It's a necessary part of good medicine, even in superbly "optimized" systems. Operations like that lap chole have taught me how easily error can occur, but they've also showed me something else: effort does matter; diligence and attention to the minutest details can save you.

This may explain why many doctors take exception to talk of "systems problems," "continuous quality improvement," and "process reëngineering." It is the dry language of structures, not people. I'm no exception: something in me, too, demands an acknowledgment of my autonomy, which is also to say my ultimate culpability. Go back to that Friday night in the E.R., to the moment when I stood, knife in hand, over Louise Williams, her lips blue, her throat a swollen, bloody, and suddenly closed passage. A systems engineer might have proposed some useful changes. Perhaps a backup suction device should always be at hand, and better light more easily available. Perhaps the institution could have trained me better for such crises, could have required me to have operated on a few more goats. Perhaps emergency tracheostomies are so difficult under any circumstances that an automated device could have been designed to do a better job. But the could-haves are infinite, aren't they? Maybe Williams could have worn her seat belt, or had one less beer that night. We could call any or all of these factors latent errors, accidents waiting to happen.

But although they put the odds against me, it wasn't as if I had no chance of succeeding. Good doctoring is all about making the most of the hand you're dealt, and I failed to do so. The indisputable fact was that I hadn't called for help when I could have, and when I plunged the knife into her neck and made my horizontal slash my best was not good enough. It was just luck, hers and mine, that Dr. O'Connor somehow got a breathing tube into her in time.

There are all sorts of reasons that it would be wrong to take my license away or to take me to court. These reasons do not absolve me. Whatever the limits of the M. & M., its fierce ethic of personal responsibility for errors is a formidable virtue. No matter what measures are taken, medicine will sometimes falter, and it isn't reasonable to ask that it achieve perfection. What's reasonable is to ask that medicine never cease to aim for it.

George Johnson

Of Mice and Elephants: A Matter of Scale

FROM THE *NEW YORK TIMES*

An ant carrying a twig is hoisting many times its own body weight. Why then can't people pick up ponderosa pines? Biologists call this the scaling problem. George Johnson, who has made a distinguished career of covering scientists and science on the frontiers of physics, mathematics, cognitive science, and complexity, reports on a pioneering partnership that has come up with a fresh approach to this puzzle. Studying the patterns that emerge at the borderline between order and chaos, a physicist and two biologists from the Santa Fe Institute have joined forces to develop a model that sheds new light on the phenomenon.

Scientists, intent on categorizing everything around them, sometimes divide themselves into the lumpers and the splitters. The lumpers, many of whom flock to the unifying field of theoretical physics, search for hidden laws uniting the most seemingly diverse phenomena: Blur your vision a little and lightning bolts and static cling are really the same thing. The splitters, often drawn to the biological sciences, are more taken with diversity, reveling in the 34,000 variations on the theme spider, or the 550 species of coniferous trees.

But there are exceptions to the rule. When two biologists and a physicist, all three of the lumper persuasion, recently joined forces at the Santa Fe Institute, an interdisciplinary research center in northern New Mexico, the result was an advance in a problem that has bothered scientists for decades: the origin of biological scaling. How is one to explain the subtle ways in which various characteristics of living creatures—their life spans, their pulse rates, how fast they burn energy—change according to their body size?

As animals get bigger, from tiny shrew to huge blue whale, pulse rates slow down and life spans stretch out longer, conspiring so that the number of heartbeats during an average stay on Earth tends to be roughly the same, around a billion. A mouse just uses them up more quickly than an elephant.

Mysteriously, these and a large variety of other phenomena change with body size according to a precise mathematical principle called quarter-power scaling. A cat, 100 times more massive than a mouse, lives 100 to the one-quarter power, or about three times, longer. (To calculate this number, take the square root of 100, which is 10, and then take the square root of 10, which is 3.2.) Heartbeat scales to mass to the minus one-quarter power. A cat's heart thus beats a third as fast as a mouse's.

The Santa Fe Institute collaborators—Dr. Geoffrey West, a physicist at Los Alamos National Laboratory, and two biologists at the University of New Mexico, Dr. Jim Brown and Dr. Brian Enquist—have drawn on their different kinds of expertise to propose a model for what causes certain kinds of quarter-power scaling, which they have extended to the plant kingdom as well.

In their theory, scaling emerges from the geometrical properties of the internal networks animals and plants use to distribute nutrients. But almost as interesting as the details of the model is the collaboration itself. It is rare enough for scientists of such different persuasions to come together, rarer still that the result is hailed as an important development.

"Scaling is interesting because, aside from natural selection, it is one of the few laws we really have in biology," said Dr. John Gittleman, an evolutionary biologist at the University of Virginia. "What is so elegant is that the work makes very clear predictions about causal mechanisms. That's what had been missing in the field."

Dr. Brown said: "None of us could have done it by himself. It is one of the most exciting things I've been involved in."

It might seem that because a cat is a hundred times more massive than a mouse, its metabolic rate, the intensity with which it burns energy, would be a hundred times greater—what mathematicians call a linear relationship. After all, the cat has a hundred times more cells to feed.

But if this were so, the animal would quickly be consumed by a fit of spontaneous feline combustion, or at least a very bad fever. The reason: the surface area a creature uses to dissipate the heat of the metabolic fires does not grow as fast as its body mass. To see this, consider (like a good lumper) a mouse as an approximation of a small sphere. As the sphere grows larger, to cat size, the surface area increases along two dimensions but the volume increases along three dimensions. The size of the biological radiator cannot possibly keep up with the size of the metabolic engine.

If this was the only factor involved, metabolic rate would scale to body mass to the two-thirds power, more slowly than in a simple one-to-one relationship. The cat's metabolic rate would be not 100 times greater than the mouse's but 100 to the power of two-thirds, or about 21.5 times greater.

But biologists, beginning with Max Kleiber in the early 1930s, found that the situation was much more complex. For an amazing range of creatures, spanning in size from bacteria to blue whales, metabolic rate scales with body mass not to the two-thirds power but slightly faster—to the three-quarter power. Evolution seems to have found a way to overcome in part the limitations imposed by pure geometric scaling, the fact that surface area grows more slowly than size. For decades no one could plausibly say why.

Kleiber's law means that a cat's metabolic rate is not a hundred or 21.5 times greater than a mouse's, but about 31.6—100 to the three-quarter power. This relationship seems to hold across the animal kingdom, from shrew to blue whale, and it has since been extended all the way down to single-celled organisms, and possibly within the cells themselves to the internal structures called mitochondria that turn nutrients into energy.

The Scientists: Common Simplicity Starts to Emerge

LONG BEFORE MEETING Dr. Brown and Dr. Enquist, Dr. West was interested in how scaling manifests itself in the world of subatomic particles. The strong nuclear force, which binds quarks into neutrons, protons, and other particles, is weaker, paradoxically, when the quarks are closer together, but stronger as they are pulled farther apart—the opposite of what happens with gravity or electromagnetism. Scaling also shows up in Heisenberg's uncertainty principle: the more finely you measure the position of a particle, viewing it on a smaller and smaller scale, the more uncertain its momentum becomes.

"Everything around us is scale-dependent," Dr. West said. "It's woven into the fabric of the universe."

The lesson he took away from this was that you cannot just naively scale things up. He liked to illustrate the idea with Superman. In two panels labeled "A Scientific Explanation of Clark Kent's Amazing Strength," from Superman's first comic book adventure in 1938, the artists invoked a scaling law: "The lowly ant can support weights hundreds of times its own. The grasshopper leaps what to man would be the space of several city blocks." The implication was that on the planet Krypton, Superman's home, strength scaled to body mass in a simple linear manner: If an ant could carry a twig, a Superman or Superwoman could carry a giant ponderosa pine.

But in the rest of the universe, the scaling is actually much slower. Body mass increases along three dimensions, but the strength of legs and arms, which is proportional to their cross-sectional area, increases along just two dimensions. If a man is a million times more massive than an ant, he will be only 1,000,000 to the two-thirds power stronger: about 10,000 times, allowing him to lift objects weighing up to a hundred pounds, not thousands.

Things behave differently at different scales, but there are orderly ways—scaling laws—that connect one realm to another. "I found this enormously exciting," Dr. West said. "That's what got me thinking about scaling in biology."

At some point he ran across Kleiber's law. "It is truly amazing because life is easily the most complex of complex systems," Dr. West said. "But in spite of this, it has this absurdly simple scaling law. Something universal is going on."

Brian Enquist became hooked on scaling as a student at Colorado College in Colorado Springs in 1988. When he was looking for a graduate school to study ecology, he chose the University of New Mexico in Albuquerque partly because a professor there, Dr. Jim Brown, specialized in how scaling occurred in ecosystems.

There are obviously very few large species, like elephants and whales, and a countless number of small species. But who would have expected, as Dr. Enquist learned in one of Dr. Brown's classes, that if one drew a graph with the size of animals on one axis and the number of species on the other axis, the slope of the resulting line would reveal another quarter-power scaling law? Population density, the average number of offspring, the time until reproduction—all are dependent on body size scaled to quarter-powers.

"As an ecologist you are used to dealing with complexity—you're essentially embedded in it," Dr. Enquist said. "But all these quarter-power scaling laws hinted that something very general and simple was going on."

The examples Dr. Brown had given all involved mammals. "Has anyone found similar laws with plants?" Dr. Enquist asked. Dr. Brown said: "I have no idea. Why don't you find out?"

After sifting through piles of data compiled over the years in agricultural and forestry reports, Dr. Enquist found that the same kinds of quarter-power scaling happened in the plant world. He even uncovered an equivalent to Kleiber's law.

It was surprising enough that these laws held among all kinds of animals. That they seemed to apply to plants as well was astonishing. What was the common mechanism involved? "I asked Jim whether or not we could figure it out," Dr. Enquist recalled. "He kind of rubbed his head and said, 'Do you know how long this is going to take?' "

They assumed that Kleiber's law, and maybe the other scaling relationships, arose because of the mathematical nature of the networks both animals and trees used to transport nutrients to all their cells and carry away the wastes. A silhouette of the human circulatory system and of the roots and branches of a tree look remarkably similar. But they knew that precisely modeling the systems would require some very difficult mathematics and physics. And they wanted to talk to someone who was used to trafficking in the idea of general laws.

"Physicists tend to look for universals and invariants whereas biologists often get preoccupied with all the variations in nature," Dr. Brown said. He knew that the Santa Fe Institute had been established to encourage broad-ranging collaborations. He asked Mike Simmons, then an institute administrator, whether he knew of a physicist interested in tackling biological scaling laws.

The Collaboration: Learning to Speak New Languages

DR. WEST LIKED TO JOKE that if Galileo had been a biologist, he would have written volumes cataloging how objects of different shapes fall from the Leaning Tower of Pisa at slightly different velocities. He would not have seen through the distracting details to the underlying truth: If you ignore air resistance, all objects fall at the same rate regardless of their weight.

But at their first meeting in Santa Fe, he was impressed that Dr. Brown and Dr. Enquist were interested in big, all-embracing theories. And they were impressed that Dr. West seemed like a biologist at heart. He wanted to know how life worked.

It took them a while to learn each other's languages, but before long they were meeting every week at the Santa Fe Institute. Dr. West would show the biologists how to translate the qualitative ideas of biology into precise equations.

And Dr. Brown and Dr. Enquist would make sure Dr. West was true to the biology. Sometimes he would show up with a neat model, a physicist's dream. No, Dr. Brown and Dr. Enquist would tell him, real organisms do not work that way.

"When collaborating across that wide a gulf of disciplines, you're never going to learn everything the collaborator knows," Dr. Brown said. "You have to develop an implicit trust in the quality of their science. On the other hand, you learn enough to be sure there are not miscommunications."

They started by assuming that the nutrient supply networks in both animals and plants worked according to three basic principles: the networks branched to reach every part of the organism and the ends of the branches (the capillaries and their botanical equivalent) were all about the same size. After all, whatever the species, the sizes of cells being fed were all roughly equivalent. Finally, they assumed that evolution would have tuned the systems to work in the most efficient possible manner.

What emerged closely approximated a so-called fractal network, in which each tiny part is a replica of the whole. Magnify the network of blood vessels in a hand and the image resembles one of an entire circulatory system. And to be as efficient as possible, the network also had to be "area-preserving." If a branch split into three daughter branches, their cross-sectional areas had to add up to that of the parent branch. This would insure that blood or sap would continue to move at the same speed throughout the organism.

The scientists were delighted to see that the model gave rise to three-quarter-power scaling between metabolic rate and body mass. But the system worked only for plants. "We worked through the model and made clear predictions about mammals," Dr. Brown said, "every single one of which was wrong."

In making the model as simple as possible, the scientists had hoped they could ignore the fact that blood is pumped by the heart in pulses and treat mammals as though they were trees. After studying hydrodynamics, they realized they needed a way to slow the pulsing blood as the vessels got tinier and tinier. These finer parts of the network would not be area-preserving but area-increasing: the cross sections of the daughter branches would add up to a sum greater than the parent branch, spreading the blood over a larger area.

After adding these and other complications, they found that the model also predicted three-quarter-power scaling in mammals. Other quarter-power scaling laws also emerged naturally from the equations. Evolution, it seemed, has overcome the natural limitations of simple geometric scaling by developing these very efficient fractal-like webs.

The Model: Strange Prediction Comes to Pass

SOMETIMES IT ALL SEEMED too good to be true. One Friday night, Dr. West was at home playing with the equations when he realized to his chagrin that the model predicted that all mammals must have about the same blood pressure. That could not be right, he thought. After a restless weekend, he called Dr. Brown, who told him that indeed this was so.

The model was revealed, about two years after the collaboration began, on April 4, 1997, in an article in *Science*. A follow-up last fall in *Nature* extended the ideas further into the plant world.

More recently the three collaborators have been puzzling over the fact that a version of Kleiber's law also seems to apply to single cells and even to the energy-burning mitochondria inside cells. They assume this is because the mitochondria inside the cytoplasm and even the respiratory components inside the mitochondria are arranged in fractal-like networks.

For all the excitement the model has caused, there are still skeptics. A paper published last year in *American Naturalist* by two scientists in Poland, Dr. Jan Kozlowski and Dr. January Weiner, suggests the possibility that quarter-power scaling across species could be nothing more than a statistical illusion. And biologists persist in confronting the collaborators with single species in which quarter-power scaling laws do not seem to hold.

Dr. West is not too bothered by these seeming exceptions. The history of physics is replete with cases where an elegant model came up against some recalcitrant data, and the model eventually won. He is now working with other collaborators to see whether river systems, which look remarkably like circulatory systems, and even the hierarchical structure of corporations obey the same kind of scaling laws.

The overarching lesson, Dr. West says, is that as organisms grow in size they become more efficient. "That is why nature has evolved large animals," he said. "It's a much better way of utilizing energy. This might also explain the drive for corporations to merge. Small may be beautiful, but it is more efficient to be big."

Jonathan Weiner

Lord of the Flies

FROM *THE NEW YORKER*

Is there a "shyness gene"? A "thrill-seeking gene"? A "gay gene"? In the past few years scientists have made headlines by announcing how all sorts of behaviors can be linked to specific genes. While many of these claims are overblown, the methods behind them are based on sound science. And one of the most important pioneers of this science is the formidable Seymour Benzer, whose lifetime study of fruit flies has resulted in several major gene identifications. Jonathan Weiner's profile of Benzer is a colorful portrait of a brilliant eccentric, captured in the wake of another important discovery by his world-famous lab.

The last time I saw the biologist Seymour Benzer, he was hiding from reporters in his office at the California Institute of Technology, in Pasadena. His laboratory runs along two corridors on the second floor of Church Hall, and his office is at the corner where the corridors meet. When I walked in, he was on the phone arranging for Caltech's public-relations office to block his outside calls. "The *methuselah* thing," he told me. Benzer had just published a report, in *Science,* about a mutant fruit fly that lives for approximately a hundred days—about forty days longer than a normal fly. Now science reporters were calling Benzer as if he had discovered the Fountain of Youth. Since the year was almost over, I asked him if he was going away for the holidays. "I don't want to travel right now," he said. "To me, getting back into the lab would be enough of an adventure."

Benzer, who is seventy-seven years old, has always tried to keep himself far from the public eye. He started out, in the early forties, as a physicist, and his experiments with semiconductors helped lead to the invention of the transistor. Working as a biologist, in the fifties, he made the world's first detailed map of the interior of a gene, which helped launch the science of molecular biology. And a research program he began at his laboratory in Pasadena in the sixties—a program that is, like Benzer himself, both whimsical and serious—is now central to the study of genes and behavior. He is one of the least known and least vain of the century's great scientists, with a genius for simple experiments that open new fields.

Late one night a few years ago, Benzer, who keeps an owl's hours, led me into the laboratory next to his office and showed me his first behavior experiment, which he had performed at Caltech in 1965. Since then, the room hadn't changed much. One wall held little bins that Benzer had labelled in a spidery black script: "Lenses," "Mirrors," "Toothpicks," "Pipe Cleaners"—anything for an experiment in the middle of the night, including "Teeth (Human and Shark)."

As I watched, Benzer laid a fifteen-watt light bulb on a bench top and turned it on. Next to it he put a half-pint milk bottle made of heavy, scratched glass with scrolls of antique advertising ("5¢—Just a Little Better") and a foam-rubber plug. Inside the bottle was a madding crowd of *Drosophila melanogaster*—fruit flies. Benzer unplugged the bottle and dumped the flies into an apparatus that he had designed in the sixties, a collection of test tubes bound together in molded plastic. He turned out the overhead lights, put on his reading glasses, and leaned forward to see what the flies would do.

Most of the flies set out through the glass tunnels toward the glow of the light bulb—some running, some walking, some meandering. A few stayed put. Benzer repeated the experiment five times, and the flies, by their own journeys through the test tubes, sorted themselves neatly into groups. Flies that chose the light every time reached Tube 5. Flies that chose the light four times ended up in Tube 4. One lone fly that never chose the light remained right where it started, in Tube Zero. Benzer, who looks unusually young and energetic for his age but whose manner is sometimes modest to the point of self-effacement, watched without expression. "O.K., there we are" was all he said as he switched on the overhead lights.

———

SEPARATING THE LIGHT-LOVING FLIES from the dark-loving flies had launched Benzer on his greatest scientific adventure: a study of the ways that differences in genes can lead to differences in behavior. He was fascinated by the infinite variety of human nature, and he decided to begin his quest by examining the far more primitive behavior of *Drosophila,* an insect that is not much bigger than an asterisk. Using his light-at-the-end-of-the-tunnel device and other gadgets, Benzer searched for flies with aberrant instincts—flies with normal-looking bodies but bizarre lives. In the years that followed, he and his students discovered an astonishing number of genes that may illuminate aspects of human nature. Their findings include everything from the *methuselah* gene—which may help biologists understand what sets the span of human life—to genes involved in sleep disorders, memory problems, and neurodegenerative diseases. Indeed, so many fly genes have turned out to be like ours that one fly researcher calls flies "little people with wings."

Benzer's pioneering work with flies has put him at the center of one of the most controversial issues in science: the extent to which our genes shape who we are and what we do. The issue has been most sharply joined at Harvard, where two of the world's leading evolutionary biologists—Richard Lewontin and Edward O. Wilson—have taken opposing stands. Lewontin, who in 1984 co-wrote an influential polemic called *Not in Our Genes,* believes that studies of the genes of flies, worms, mice, or even human beings, don't tell us much about human behavior. He thinks we are essentially creatures of culture, not instinct. Wilson, on the other hand, told me he believes that the study of genes and behavior is beginning to revitalize the battered field of psychology—indeed, that it will one day enable us to understand human nature from the ground up. "Better Benzer than Freud!" Wilson said. "Quote me. Better Benzer than Freud!"

Although Benzer refuses to enter the debate, he keeps a human brain in a plastic bucket of formaldehyde in his lab—a reminder of his conviction that the genes in his fruit flies do relate to the genes in the bucket. "Psychology," he tells me in his low-key way, "is going to be changed."

THE MODERN THEORY of the gene came out of the first fly room—No. 613 Schermerhorn Hall, at Columbia University. There, in the early decades of this century, the biologist Thomas Hunt Morgan bred mutant fruit flies—flies with yellow bodies instead of brown ones, vermillion eyes instead of red ones, curly wings instead of straight ones—and he watched each trait pass down through generations of flies, in hundreds and hundreds of half-pint milk bottles. Morgan was repeating and extending the work that the Austrian monk Gregor Mendel had done in the eighteen-fifties with garden peas. Mendel had bred strains of plants with smooth peas and wrinkled peas, with short stems and tall stems, and so on. He had concluded that each trait in every living creature is governed by a separate factor. But Mendel did not know what—or where—those factors might be.

More than half a century later, Morgan found part of the answer. He discovered that genes are situated on tiny threads called chromosomes, which can be seen, through a microscope, in the nucleus of every egg, in the head of every spermatozoon, and in every living cell. Morgan named the first gene he found *white*, because he had discovered it in a mutant fly with white eyes. He and his students, who became known as "Morgan's Raiders," traced the *white* gene to a microscopic point on the fly's X chromosome, with *yellow* just to the left of it and *vermillion* just to the right. It was the first gene map, and it was a simple one—just a few points on a straight line.

Morgan's research had a certain madcap tone—a tone that Benzer later revived. In Morgan's fly room, escaped flies drifted around a garbage can that was never completely emptied. Other flies swarmed around a bunch of bananas that hung in a corner. The Raiders arrived for work each morning bearing empty milk bottles that—legend has it—they had stolen from Manhattan stoops. "Morgan was a bit crazy," one of Benzer's former students, the drosophilist Jeff Hall, told me. "He used to say, 'To know your organism, you must eat it.' Not just the flies but also the pupae. And not just to horrify people but to *know*. When people stared at him, he said, 'They taste like Grape-Nuts.' "

Eventually, Morgan moved his operation to Caltech, which had become one of the world's great research centers for chemistry and physics. In a Nobel Prize lecture, in 1934, he noted that his genetic maps were purely schematic. He still wanted to know what genes are made of, and how they work, and he hoped that physicists would tell him.

THAT SAME YEAR, in Bensonhurst, Brooklyn, Seymour Benzer was given a microscope as a bar-mitzvah present. He took it down to the family basement, where he had built a laboratory. There he performed what seemed to him the obvious first experiment: he used his new microscope to survey hundreds of thousands of thrashing tadpoles with tiny heads—his sperm.

No one else in the Benzer family was interested in science. His parents had come to America from a shtetl west of Warsaw; in New York, they worked in the needle trades. But Seymour was the only boy in the family—he was the prince, "the egg with two yellows," to use an Old World expression—and on most afternoons and evenings his parents and his three sisters left him free to play stickball on the street or to putter in the basement, striking mad-scientist poses and staring through his microscope at deconstructed houseflies. At fifteen, he won a Regents Scholarship and entered Brooklyn College. In those days, the typical introductory biology course was closer to natural history than to physics and chemistry: it was a survey of all the plants and animals on earth. Benzer asked to skip to more advanced classes, but that was against the rules of the biology department. "So," he recalls, "being a stupid, pigheaded, cocky young guy, I told them the hell with it, and didn't take any biology at all."

Instead, he studied physics and chemistry. In 1942, just as the United States was entering the Second World War, he enrolled as a graduate student in the physics department at Purdue University, in Indiana. There, in a secret wartime laboratory, he and a small team of physicists worked on radar technology. At the time, radar sets used rectifiers made of silicon, a semiconductor. Because these rectifiers were unreliable—when high voltage hit them, they burned out—Purdue's team was trying to replace silicon with germanium, another semiconductor. Benzer discovered a germanium crystal that could withstand extremely high voltages. After the war, the crystal was used at Bell Labs to build the first transistor.

In 1946, a friend at Purdue gave Benzer a copy of *What Is Life?*, a little book by the Austrian quantum physicist Erwin Schrödinger. The book contained speculations on the physical nature of the gene by the German quantum physicist Max Delbrück. For Benzer and a whole generation of other young scientists—including James Watson, in Chicago, and Francis Crick, in London—the experience of reading this slender volume turned the gene into the great unsolved mystery of biology. When Benzer learned that Delbrück, who was now at Caltech, had organized a summer course at the Cold Spring Harbor Laboratory, on Long Island, he signed up.

"Within one day at Cold Spring Harbor," Benzer later recalled, "I became committed to being a biologist." After he and his wife, Dotty, returned to Purdue in the fall, he announced that he was changing fields. "People thought I was nuts," he says. "Here it was, the semiconductor thing was booming. My boss, Karl Lark-Horovitz, and I had got six patents on the work we had done with the semiconductor." Most of Benzer's Purdue colleagues were planning to form electronics companies and get rich. But Benzer told them matter-of-factly, "I'm interested in biology."

IN HIS EARLY YEARS as a biologist, Benzer lived like a gypsy. He worked with Delbrück at Caltech, then with François Jacob and Jacques Monod at the Pasteur Institute, in Paris. In the fall of 1953—the year that Watson and Crick put together their double-helix model of the spiralling molecule of DNA, which is the main constituent of every chromosome—he was back in Indiana, in his lab at Purdue. Although Benzer was nominally a professor of physics, he had taken to spending most of his time experimenting with viruses and bacteria. He was not trying to understand the viruses and bacteria themselves; rather, he was using them as tools to discover the physical nature of the gene, studying heredity in the simplest forms of life. Flies are tiny, but viruses are submicroscopic. Benzer later wrote that in a test tube, in twenty minutes, he could conduct "an experiment yielding a quantity of genetic data that would require, if humans were used, the entire population of the earth." One day that fall, he realized that by manipulating virus particles he could, in a sense, get inside the gene. The result would be a manifold expansion of Morgan's genetic maps—like magnifying a painting by Seurat until he could map the terrain of a single dot. Among other things, a map as detailed as that would help to confirm that Watson and Crick were right about the double helix.

After much thought, Benzer wrote out a sketch of his plans. During a meeting in Amsterdam in 1954, he showed the sketch to his old mentor, Max Delbrück. Delbrück found it outrageous. Benzer cherishes the comments that Delbrück scribbled on his manuscript: "You must have drunk a triple highball before writing this. This is going to be offensive to a lot of people that I respect."

Actually, like all of Benzer's experiments, this one was ingeniously simple. He focussed on a chromosome region called rII, and over the next ten years he discovered so much about it that his map grew into a mural that stretched across his lab. It was the first fine-scale map of a gene's interior. Biologists of a certain age can still remember the impression that Benzer made with his map at conferences in the fifties, carrying it onstage and unrolling it like a Torah scroll. Today, the map work that Benzer began has become even more detailed—indeed, has reached the bedrock molecular level—and is the single biggest project in the history of biology. International teams of molecular biologists are competing to map and read every fly gene and every human gene, at a cost of billions of dollars and at rates of more than a hundred million new letters of genetic code a year.

Benzer's reputation as a pioneer in biology was enhanced among his friends by his eccentricities. He always wore many layers of clothing; he did his best work in the middle of the night; and he amazed them by the things he ate. François Jacob, of the Pasteur Institute, describes Benzer in his memoirs: "Every day, at lunch, he brought some unusual dish—cow's udder, bull's testicles, crocodile tail, filet of snake—which he had unearthed on the other side of Paris and which he simmered on his Bunsen burner." Benzer ate like that at home, too, with the indulgence of Dotty, who watched as he sampled caterpillars, ducks' feet, horse meat, and snails. One morning in Paris, their daughter Barbie woke up with her eyes swollen shut. When the doctor asked, "Has she eaten anything unusual lately?" Benzer was almost too mortified to be truthful.

BY THE EARLY SIXTIES, molecular biologists had learned so much about the gene that they doubted whether they would ever encounter any further mysteries to equal the ones they had dispelled. Benzer says, "It was a little bit like the physicists at the end of the nineteenth century saying, 'All we have left to do is find one more decimal place.' "

Shortly after his fortieth birthday, Benzer's curiosity took another turn. He, Dotty, and their two young daughters, Barbie and Martha, often spent time in Cold Spring Harbor. At the beach, Benzer was struck by how unlike the two girls were as they played by the water's edge: Barbie was so lively, Martha was so calm. Martha had been placid since her first week in the crib. He didn't think that he and his wife could possibly have made that much difference in the temperaments of the two girls—the difference had to be in the genes. It occurred to him that whenever he read something about rII he felt bored, but whenever he read or talked about behavior and personality he felt engaged. He was using what Crick calls "the gossip test"—that is, "what you are really interested in is what you gossip about."

So Benzer began preparing for his next career jump. He knew that scientists and philosophers had argued forever about "nature versus nurture"— generations of thinkers trapped like flies in a fly bottle, pirouetting around and around. But now they had a solid foundation from which to work—the gene. Now they could try to trace the connections between the gene and the brain, and between the brain and behavior. Benzer asked Dotty to buy brains at the butcher's shop: sheep, cow, goat, pig, and chicken. One by one, she brought them home, and one by one he dissected them, usually in the middle of the night. Afterward, he ate them.

In 1965, Benzer went to Caltech as a visiting researcher, still looking for the right animal with which to start his new study. By chance, his lab was a short distance from what had been Morgan's last fly room. Although Morgan had died in 1945, his collected works still sat in rows of file cabinets out in the hall. Morgan's favorite Raider, Alfred Sturtevant, who had followed Morgan to Caltech, was retired now, but one of his students, Ed Lewis, had inherited the old fly room, along with thousands of mutant flies and the milk bottles. One night, Benzer stopped by the fly room and borrowed a few test tubes and a fly bottle. He put a small light bulb on his bench top, turned out the overhead lights, and watched the flies run toward the light.

IN THE LATE SIXTIES and early seventies, Benzer gathered a remarkable group of graduate and postdoctoral students around him—Benzer's Raider. His friend Francis Crick told me that without Benzer's enormous success in molecular biology he could never have drawn students into such a wildly improbable research program. "If he hadn't done it," Crick says, "no one else would have done it." Benzer's Raiders worked in the middle of the night, rambling from fly room to fly room in Church Hall, and sampling—sometimes vomiting—an assortment of snacks: haggis, century eggs, fish lips, all of which Benzer insisted on sharing whenever he and his students paused for what he called "tearing a herring." William Harris, who was a graduate student of Benzer's in those early days and who is now chairman of the Department of Anatomy at Cambridge, recalls, "The whole lab was infected by his spirit of the unusual."

The light-at-the-end-of-the-tunnel apparatus was the prototype for a series of experiments in which Benzer and his Raiders looked for mutant instincts and mutant behavior among the flies. To insure that they would find plenty of mutants, they fed their specimens a mutagen—a poison called ethyl methane sulfonate—that will generally change only one letter of the genetic code at a time.

Their first breakthrough came in a study of the flies' sense of time. Naturalists had suspected for centuries that living things carry built-in clocks. Morning glories know when to bloom. Bears in caves know when to wake up. Many people can successfully order themselves to wake up at seven o'clock in the morning. Benzer, who tended to wake up around noon, had always suspected that he had a mutant clock in him—a mutation that helped him spend many nights of happy solitude in his lab.

Fruit flies are early risers. (*Drosophila* means "lover of dew.") In Benzer's fly room, they stick to their daily cycle even if they are put in a room that is dark around the clock. At the hour of their virtual sunrise, they begin moving around more or less simultaneously. They take naps in the middle of their day. They get quiet again at their virtual sundown. Fruit flies hatch—or, in the jargon, eclose—at dawn. "They're very endearing when they first come out," Benzer says. "Moist and vulnerable. And all that behavior is built in!"

In 1968, Benzer and one of his first graduate students at Caltech, Ronald J. Konopka, isolated a strain of flies that neither eclosed nor awakened at dawn. Their hours were completely unpredictable; they were clock mutants. Benzer and Konopka also found a strain of flies whose clocks ran about five hours fast, and then a strain whose clocks ran four or five hours slow. All three mutants turned out to have something wrong with a single gene, which Benzer and Konopka named *period*. At a party in Pasadena, Benzer and Konopka told Max Delbrück that they had found a mutant fly that had something wrong with its sense of time, and they had traced the gene to the fly's X chromosome.

"I don't believe it," Delbrück said.

"But, Max," Benzer said, "we found the gene!"

"I don't believe a word of it," Delbrück said.

Still, Benzer knew he was right.

He and his students also looked at flies' instincts for courtship and copulation. Put a virgin male and a virgin female fruit fly in a vial, and the action runs much the same course again and again: even if the male has never seen a female fly before, he seems to experience, as Benzer puts it, "an immediate 'Aha!' reaction." Within seconds, he maneuvers himself so that he is facing her head from one side. Then he holds out one wing toward her in a kind of salute, and sets it vibrating. Hurrying around to her other side, he holds out his other wing toward her and vibrates that: second verse, same as the first. Working in his lab at night, Benzer could just hear the flies' "*Eine Kleine Nachtmusik*" if he lowered his ear to an open vial. If a male fly sang the right song and if the female was a virgin, then the rest of the action proceeded in a distinctive series of steps that were, as Benzer puts it, "only too embarrassingly anthropomorphic." Typically, copulation lasted twenty minutes.

To begin the genetic dissection of these instincts, Benzer encouraged Jeff Hall, then a postdoctoral student, to collect flies that were luckless in love. One such mutant had turned up in a fly room at the University of California in San Diego. The male courted vigorously but never copulated. Hall named it *celibate*. From the same fly room he acquired a male mutant that disengaged after ten or twelve minutes and rarely fathered any children. Hall named him *coitus interruptus*. Then there was *stuck*. A *stuck* male has trouble withdrawing his penis after copulation. "The pair just stick together," Benzer says, "and they keep pulling against each other for hours or days on end. Sometimes they die of starvation."

In a fly bottle, normal males ignore other males. If a male sees another male approaching him with a wing out, in troubadour style, he flicks his own wings violently in rejection. But Hall acquired a strain of mutants from a fly room at Yale in which the males sang to each other. Sometimes three males— or five, or ten, or more—would form chains and follow each other around in the bottom of a petri dish in long, winding conga lines. When they had reached a certain pitch of frenzy, they would get right up on the walls of the petri dish and keep chaining for hours. The male flies never tried to copulate. They only formed these long lines and danced, sometimes sticking out one wing and singing the love song, as if they were shaking a tambourine. Hall and others managed to map the gene; they named it *fruitless*.

Meanwhile, Benzer was trying to figure out how to study events in the brain which lead to learning and memory. With another postdoc, Chip Quinn, he set out to see if fruit flies could learn. The two added scents to the light-at-the-end-of-the-tunnel apparatus. When the flies entered a tube scented with octanol—which to human noses smells vaguely like licorice—they experienced a shock of seventy volts. When they entered a tube scented with methyl-cyclohexanol—which smells, as someone in the lab put it, "a lot like tennis shoes in July"—the flies made their way toward the light at the end of the tunnel without receiving a shock. Quinn ran whole schools of flies through this teaching machine. Each mistake cost the fly fifteen seconds of seventy volts, which might have killed a human being but only ruffled the fly's bristles. After a while, when the flies encountered a fresh test tube that smelled like licorice, most of them did not go inside.

Soon after, another student of Benzer's discovered a strain of flies that had no talent for learning. These flies ran to the light, and climbed the walls; they flew, walked, courted, sang, and copulated like normal flies. But they never seemed to learn. Benzer's Raiders named the new mutant *dunce*.

IN THE SEVENTIES and eighties, Benzer's students began wandering out into the world to start fly rooms of their own. For molecular biology, the mutant fly known as *period* became a point of entry into the sense of time. In the last few years, the molecular model of the clock has grown ornate, as more mutants—with such names as *timeless* and *doubletime*—have been discovered. Often the discovery of a gene in the fly has led straight to its discovery in human beings, because key sequences in both genetic codes are very much alike. It seems that we human beings have *period* genes, and *timeless* genes, too. Moreover, our clocks apparently work so much like fly clocks that, in laboratory tests, some of our protein gears mesh perfectly with the flies' gears, like parts of the same watch—even though human beings and flies have not shared an ancestor since the Cambrian Period, six hundred million years ago.

Molecular biologists who breed fruit flies have also helped trace genetic events that decide the sex of an embryo—any embryo, from a fly's to a human being's. And in recent years Benzer's students have discovered more learning and memory genes, including *latheo,* which is named after Lethe, the river in Hades that caused souls to forget their pasts, and *linotte,* which is named after the French expression *tête de linotte,* or "birdbrain." At the moment, the most intensely studied memory gene is *creb,* which gives a fly the genetic equivalent of an on-and-off switch. A few years ago, in Cold Spring Harbor, researchers engineered a fly whose switch was jammed "on." This fly learned its lessons unnaturally fast and remembered them for a whole week. The genetic engineers had given the fly the equivalent of a photographic memory.

DOTTY BENZER DIED OF CANCER in the spring of 1978. Benzer, who was despondent, decided to drop his own work on behavior, and let his students carry on. A few of them still feel abandoned. "Well, I'd jumped before," Benzer told me, and he added, "You know the song," referring to a celebration party for Max Delbrück when he won the Nobel Prize, in 1969. On that occasion, one of Benzer's friends sang a song about him to the tune of "Jimmy Crack Corn." Three verses mapped the Benzer career:

> Physics was fun but I don't care,
> I'm on to something else next year,
> I must stick with the new frontier
> Until I'm old and gray.

> Genetics was fun but I don't care,
> I'm on to something else next year,
> I must stick with the new frontier
> Until I'm old and gray.

> Behavior was fun but I don't care,
> I'm on to something else next year,
> I must stick with the new frontier
> Until I'm old and gray.

Although the song was sung with great affection, a few of Benzer's friends sometimes wonder if all this career hopping has cost him a Nobel of his own.

In the fall of 1978, Benzer met Carol Miller, a neuropathologist at the University of Southern California School of Medicine. She was seventeen years younger than Benzer, and to him she looked like the woman in Vermeer's "Girl with a Pearl Earring." They soon discovered that they shared an obsession with the machinery of the brain and enjoyed the same sorts of novelties, such as a plastic brain with a compass in it, a plastic eyeball key chain, and a brain-shaped Jell-O mold. Benzer was dissecting fly eyes and brains; Miller was dissecting human eyes and brains. He would tell her about the bizarre mutant fly he had found that week. She would say, "That sounds like the patient I saw yesterday."

They married in 1980, and embarked on what Miller says is a true marriage of minds and brains. She adapted some of the high-tech tools he had developed in the study of defective fly brains, and used them in her research lab to study tissue samples from the brains of victims of Parkinson's, amyotrophic lateral sclerosis (Lou Gehrig's disease), Alzheimer's, and other neurodegenerative diseases. Lately, Benzer and one of his postdocs have been screening mutant flies and searching for brain-degeneration patterns that look like the ones in Miller's neuropathology textbooks. A surprising number of fly brains and human brains have problems that look alike when they're inspected under an electron microscope. Whenever Benzer finds a particularly interesting case, he tells Miller about it over dinner at their modest home in San Marino, a few minutes' drive from Caltech. The front door has a stained-glass window in which fly eyes and fly brains are disguised as flowers. A reproduction of "Girl with a Pearl Earring" hangs in the living room. They have a fourteen-year-old son, Alexander, and Benzer used to carry a picture of the boy's chromosomes in his wallet.

Occasionally, Benzer and Miller meet for lunch in his fly room. During the meal, Benzer's postdoc shows them photographs of fly brains. Then Benzer's Raiders name the latest fly mutants after the foods that the flies' brain lesions remind them of: *eggroll, popcorn, sponge cake, bubblegum, meringue, chocolate chip.*

THE RETIRED SENATOR William Proxmire used to give out Golden Fleece awards to scientists who, in his view, were wasting the taxpayers' money on ridiculous research projects. Benzer once heard a rumor that he was a Golden Fleece nominee for his fly work. Today, if Richard Lewontin were to revive the award, he might nominate Benzer, too. I visited Lewontin at Harvard's Museum of Comparative Zoology, where he studies flies not to understand their behavior but to understand evolution—the ways that the pressures of natural selection change the lines of the genetic code in flies, fish, or human beings. Over lunch, Lewontin maintained that Benzer and his students have found nothing in the fly that matters for the man and woman on the street. "Always the same story," he told me, in a world-weary tone. "You take a simple organism because it is simpler to study, but you factor out everything that's interesting in the process. In terms of *Drosophila*, in terms of the evolution of its courtship, the work is interesting. But it's like saying that someone has understood why I eat what I am eating because of smell perception in fruit flies." He smiled. His choice of this particular lunch, he said, did not originate in his genes. "I'm eating it because of my social position in this culture, and what's available, and what lunch I got to eat when I was a kid." In other words, this was a lunch that was not typical of his species. "Most people in the world," he said, "don't eat pizza and cookies."

One flight up in the same building, Benzer's great defender, Edward O. Wilson, told me that he finds Lewontin's argument irrelevant. "Science consists substantially of finding an entry point," he said. "And even a small advance following a breakthrough should be regarded as science of the first class. What's unreasonable is then to demand that the people who are doing the very early work come up with a whole explanation of everything. That's crazy." In Wilson's recent book *Consilience: The Unity of Knowledge,* he argues that science is beginning to piece together one "consilient" or interconnected picture of nature that includes everything from dead atoms to warm, living flesh. He sees no reason to doubt that, through the sort of work that Benzer is doing, we will one day be able to extend the reach of scientific explanation to the behavior of even the most complex organisms, including human beings.

For all of Benzer's open-field explorations, he has always been a skeptical scientist. In Church Hall, he keeps a file of headlines pertaining to genes and behavior, so that if their claims are discredited he can use them in his lectures as cautionary tales. Over the last three decades, scientists have claimed to have found in human beings genetic links to all sorts of aberrant behavior—everything from a propensity for violence to reading disabilities, manic depression, psychosis, alcoholism, autism, drug addiction, gambling addiction, attention-deficit disorder, post-traumatic stress disorder, and Tourette's syndrome. Many of those studies have turned out to be shaky, and some have been retracted.

Today, as the human genome is mapped and deciphered, more and more molecular biologists are gambling that the time is right to plunge in. Dean Hamer, of the National Institutes of Health, is one of the best known of the molecular biologists who have begun looking at the links between genes and behavior in human beings. Since 1993, Hamer's studies have produced headlines announcing the discovery of a "gay gene," a "novelty gene," and a "happiness gene."

Benzer knows that this is a field he helped to pioneer, but, like many other molecular biologists, he finds the headlines overblown and oversimplified. He is convinced that as the picture of genes and their relation to behavior becomes more complete, such traits will prove to be at least as complicated as a fly's tendency to move toward light. He has now dissected that trait into hundreds of genes and components, going from the retina deep into the brain. Even though a fly brain is so small that it is hard to see without a magnifying glass, it contains more than a hundred thousand neurons. Why, Benzer asks, should we imagine that our turnings are less complicated than a fly's?

"We have a dog, a golden retriever named Cassandra," Benzer told me recently. "Her behavior is right on target—such sweetness bred into the breed!" Still, every time he takes Cassandra for a walk along the palm-lined side streets of San Marino, he can't help wondering at all the ways in which the dog's genes and her environment interact. "The only way to tease genes and environment apart," he says, "is to keep the environment constant and change the genes. But I never say that it's *all* in the genes. Genetics can be used as a tool to attack the problem of nature versus nurture—with proper control. Nowadays, though, the pendulum has swung too far in the direction of nature. You pick out a dog based on breed, but then you sure as hell try to train it."

A FEW DAYS after Benzer's seventy-seventh birthday, last October, he and his latest crew of postdocs announced their discovery of *methuselah*—the mutant fly that lives a hundred days. Now Benzer is using *methuselah* to search for a sort of clock of clocks, whose hands can help set each lifespan. "Time again— time revisited," he says of his latest inquiry, which he expects to pursue well into the next century.

In his *Philosophical Investigations,* in 1953, Ludwig Wittgenstein asked, "What is your aim in philosophy?" He answered the question himself: "To show the fly the way out of the fly bottle." Well, the fly is out of the bottle now. When Benzer switched from physics to genetics, he thought of a fly as simply a particle, an atom of behavior. Now that he has spent a third of a century getting inside the brain of the fly, he would not dream of calling it merely an atom of behavior.

His research on *methuselah* interests him so much that sometimes he would rather work than sleep. In the middle of the night in Church Hall, he unhoods a microscope to examine the head of a fly. "About the size of the head of a pin," he says. "A hundred thousand angels on it. Dancing."

SHERYL GAY STOLBERG

The Biotech Death of Jesse Gelsinger

FROM THE *NEW YORK TIMES MAGAZINE*

By early 2000, gene therapy—using healthy genes to replace faulty ones—was at the center of an explosive public-health debate. After the death of an eighteen-year-old volunteer, Jesse Gelsinger, during the clinical trial of a new gene treatment, federal agencies placed new scrutiny on gene therapy, and the U.S. Congress launched an investigation into the safety of this potentially promising but still unproven technique. Sheryl Gay Stolberg, whose reporting for the New York Times *has helped shape public opinion on this issue, tells the story of how the search for a scientific breakthrough led to an unforeseen tragedy.*

The jagged peak of Mount Wrightson towers 9,450 feet above Tucson, overlooking a deep gorge where the prickly pear cactus that dots the desert floor gives way to a lush forest of ponderosa pine. It is said that this is as close to heaven as you can get in southern Arizona. Jesse Gelsinger loved this place. So it was here, on a clear Sunday afternoon in early November, that Paul Gelsinger laid his eighteen-year-old son to rest, seven weeks after a gene-therapy experiment cost him his life.

The ceremony was simple and impromptu. Two dozen mourners—Jesse's father; his mother, Pattie; his stepmother, Mickie; and two sisters, a brother, three doctors and a smattering of friends—trudged five miles along a steep trail to reach the rocky outcropping at the top. There, Paul Gelsinger shared stories of his son, who loved motorcycles and professional wrestling and was, to his father's irritation, distinctly lacking in ambition. Jesse was the kind of kid who kept $10.10 in his bank account. "You need $10 to keep it open," Gelsinger explained but those assembled on the mountaintop agreed that he had a sharp wit and a sensitive heart.

At Gelsinger's request, the hikers had carried Jesse's medicine bottles filled with his ashes, and now they were gathered at the edge of the peak. Steve Raper, the surgeon who gave Jesse what turned out to be a lethal injection of new genes, pulled a small blue book of poetry from his pocket. "Here rests his head upon the lap of Earth," Raper read, reciting a passage from an elegy by Thomas Gray, "a youth to Fortune and Fame unknown. / Fair Science frowned not on his humble birth." Then the surgeon, the grieving father and the rest scattered Jesse's ashes into the canyon, where they rose on a gust of wind and fell again in a powerful cloud of fine gray dust. "I will look to you here often, Jess," Paul Gelsinger said sadly.

Jesse Gelsinger was not sick before he died. He suffered from ornithine transcarbamylase (OTC) deficiency, a rare metabolic disorder, but it was controlled with a low-protein diet and drugs, thirty-two pills a day. He knew when he signed up for the experiment at the University of Pennsylvania that he would not benefit; the study was to test the safety of a treatment for babies with a fatal form of his disorder. Still, it offered hope, the promise that someday Jesse might be rid of the cumbersome medications and diet so restrictive that half a hot dog was a treat. "What's the worst that can happen to me?" he told a friend shortly before he left for the Penn hospital, in Philadelphia. "I die, and it's for the babies."

As far as government officials know Jesse's death on September 17 was the first directly related to gene therapy. The official cause, as listed on the death certificate filed by Raper, was adult respiratory distress syndrome: his lungs shut down. The truth is more complicated. Jesse's therapy consisted of an infusion of corrective genes, encased in a dose of weakened cold virus, adenovirus, which functioned as what scientists call a vector. Vectors are like taxicabs that drive healthy DNA into cells; viruses, whose sole purpose is to get inside cells and infect them, make useful vectors. The Penn researchers had tested their vector, at the same dose Jesse got, in mice, monkeys, baboons and one human patient, and had seen expected, flulike side effects, along with some mild liver inflammation, which disappeared on its own. When Jesse got the vector, he suffered a chain reaction that the testing had not predicted—

jaundice, a blood-clotting disorder, kidney failure, lung failure and brain death: in Raper's words, "multiple-organ-system failure." The doctors are still investigating; their current hypothesis is that the adenovirus triggered an overwhelming inflammatory reaction—in essence, an immune-system revolt. What they do not understand yet is why.

Every realm of medicine has its defining moment, often with a human face attached. Polio had Jonas Salk. In vitro fertilization had Louise Brown, the world's first test-tube baby. Transplant surgery had Barney Clark, the Seattle dentist with the artificial heart. AIDS had Magic Johnson. Now gene therapy has Jesse Gelsinger.

Until Jesse died, gene therapy was a promising idea that had so far failed to deliver. As scientists map the human genome, they are literally tripping over mutations that cause rare genetic disorders, including OTC deficiency, Jesse's disease. The initial goal was simple: to cure, or prevent, these illnesses by replacing defective genes with healthy ones. Biotech companies have poured millions into research—not for rare hereditary disorders but for big-profit illnesses like cancer, heart disease and AIDS. As of August, the government had reviewed 331 gene-therapy protocols involving more than 4,000 patients. Just 41 were for the "monogeneic," or single-gene, defect diseases whose patients so desperately hoped gene therapy would be their salvation.

At the same time, the science has progressed slowly; researchers have had trouble devising vectors that can carry genes to the right cells and get them to work once they are there. Four years ago, Dr. Harold Varmus, the director of the National Institutes of Health, commissioned a highly critical report about gene therapy chiding investigators for creating "the mistaken and widespread perception of success." Since then, there have been some accomplishments: a team at Tufts University has used gene therapy to grow new blood vessels for heart disease patients, for instance. But so far, gene therapy has not cured anyone. As Ruth Macklin, a bioethicist and member of the Recombinant DNA Advisory Committee, the National Institutes of Health panel that oversees gene-therapy research, says, bluntly, "Gene therapy is not yet therapy."

On December 8, the "RAC," as the committee is called, will begin a public inquiry into Jesse's death, as well as the safety of adenovirus, which has been used in roughly one-quarter of all gene-therapy clinical trials. The Penn scientists will report on their preliminary results, and investigators, who at the RAC's request have submitted thousands of pages of patient safety data to the committee, will discuss the side effects of adenovirus. Among them will be researchers from the Schering-Plough Corporation, which was running two experiments in advanced liver cancer patients that used methods similar to

Penn's. Enrollment in those trials was suspended by the Food and Drug Administration after Jesse's death. The company, under pressure from the RAC, has since released information showing that some patients experienced serious side effects, including changes in liver function and blood cell counts, mental confusion and nausea; two experienced minor strokes, although one had a history of them. Once all the data on adenovirus are analyzed at the December 8 meeting, the RAC may recommend restrictions on its use, which will almost certainly slow down some aspects of gene-therapy research.

The meeting will be important for another reason: it will mark an unprecedented public airing of information about the safety of gene therapy—precisely the kind of sharing the RAC has unsuccessfully sought in the past. Officials say gene therapy has claimed no lives besides Jesse's. But since his death, there have been news reports that other patients died during the course of experiments—from their diseases, as opposed to the therapy—and that the scientists involved did not report those deaths to the RAC, as is required. This has created a growing cloud of suspicion over gene therapy, raising questions about whether other scientists may have withheld information that could have prevented Jesse's death. That question cannot be answered until all the data are analyzed. But one thing is certain: four years after the field was rocked by Varmus's highly critical evaluation, it is now being rocked again, this time over an issue more fundamental than efficacy—safety. "I think it's a perilous time for gene therapy," says LeRoy Walters, a bioethicist at Georgetown University and former chairman of the RAC. "Until now, we have been able to say, 'Well, it hasn't helped many people, but at least it hasn't hurt people.' That has changed."

No ONE, PERHAPS, is more acutely aware of gene therapy's broken promise than Mark Batshaw, the pediatrician who proposed the experiment that cost Jesse Gelsinger his life.

At fifty-four, Batshaw, who left the University of Pennsylvania last year for Children's National Medical Center, in Washington, is tall and gangly with slightly stooped shoulders and a shy smile that gives him the air of an awkward schoolboy, which he once was. As a child, Batshaw struggled with hyperactivity: He didn't read until the third grade; in the fourth, his teacher grew so irritated at his constant chatter that she stuck his chair out in the hall. The experience has left him with a soft spot for developmentally disabled children, which is how he has become one of the world's foremost experts in urea-cycle disorders, among them OTC deficiency.

The urea cycle is a series of five liver enzymes that help rid the body of ammonia, a toxic breakdown product of protein. When these enzymes are missing or deficient, ammonia—"the same ammonia that you scrub your floors with," Batshaw explains—accumulates in the blood and travels to the brain, causing coma, brain damage and death. OTC deficiency is the most common urea-cycle disorder, occurring in one out of every 40,000 births. Its genetic mutation occurs on the X chromosome, so women are typically carriers, while their sons suffer the disease.

Severe OTC deficiency is, Batshaw says, "a devastating disease." Typically, newborns slip into a coma within seventy-two hours of birth. Most suffer severe brain damage. Half die in the first month, and half of the survivors die by age five. Batshaw was a young postdoctoral fellow when he met his first urea-cycle-disorder patient in 1973, correctly diagnosing the disease at a time when most other doctors had never heard of it. Within two years, he and his colleagues had devised the first treatment, a low-protein formula called keto-acid. Later, they came up with what remains standard therapy to this day: sodium benzoate, a preservative, and another type of sodium, which bind to ammonia and help eliminate it from the body.

But the therapy cannot prevent the coma that is often the first sign of OTC and ravages the affected infant. By the time Batshaw joined the faculty at Penn in 1988, he was dreaming of a cure: gene therapy. Patients were dreaming, too, says Tish Simon, former co-president of the National Urea Cycle Disorders Foundation, whose son died of OTC deficiency three years ago. "All of us saw gene therapy as the hope for the future," Simon says. "And certainly, if anybody was going to do it, it had to be Mark Batshaw."

GENE THERAPY BECAME A REALITY on September 14, 1990, in a hospital room at the National Institutes of Health, in Bethesda, Maryland, when a four-year-old girl with a severe immune-system deficiency received a thirty-minute infusion of white blood cells that had been engineered to contain copies of the gene she lacked. Rarely in modern medicine has an experiment been filled with so much hope; news of the treatment ricocheted off front pages around the world. The scientist who conducted it, Dr. W. French Anderson, quickly became known as the father of gene therapy. "We had got ourselves all hyped up," Anderson now admits, "thinking there would be rapid, quick, easy, early cures."

Among those keeping a close eye on Anderson's debut was Jim Wilson, a square-jawed, sandy-haired midwesterner who decided to follow his father's footsteps in medicine when he realized he wasn't going to make it in football. As a graduate student in biological chemistry, Wilson had taken a keen interest in rare genetic diseases. "All I did," he says, "was dream about gene therapy."

Today, as director of the Institute for Human Gene Therapy at the University of Pennsylvania, Wilson is in an excellent position to make that dream a reality. Headquartered in a century-old building amid the leafy maple trees and brick sidewalks of the picturesque Penn campus, the six-year-old institute, with 250 employees, state-of-the-art laboratories and a $25 million annual budget, is the largest academic gene-therapy program in the nation. In a field rife with big egos, Wilson is regarded as first-rate. "Present company excluded," Anderson says, "he's the best person in the field."

Batshaw was banging on Wilson's door even before Wilson arrived at Penn in March 1993, and within a month they were collaborating on studies of OTC-deficient mice. Their first task was to develop a vector. Adenovirus seemed a logical choice.

There had been some early problems with safety—a 1993 cystic fibrosis experiment was shut down when a patient was hospitalized with inflamed lungs—but Wilson and Batshaw say they figured out how to make a safer vector by deleting extra viral genes. Adenovirus was the right size: when its viral genes were excised, the OTC gene fit right in. It had a "ZIP code," on it, Batshaw says, that would carry it straight to the liver. And while its effects did not last, it worked quickly, which meant that it might be able to reverse a coma, sparing babies from brain damage. "It wasn't going to be a cure soon," Batshaw says, "but it might be a treatment soon."

The mouse experiments were encouraging. Mice that had the therapy survived for two to three months even while fed a high-protein diet. Those that lacked the treatment died. "It wasn't subtle," Wilson says. "We felt pretty compelled by that." But when the team contemplated testing in people, they ran smack into an ethical quandary: who should be their subjects?

To Wilson, the answer seemed obvious: sick babies. Arthur Caplan, the university's resident bioethics expert, thought otherwise. Caplan says parents of dying infants are incapable of giving informed consent: "They are coerced by the disease of their child." He advised Wilson to test only stable adults, either female carriers or men like Jesse, with partial enzyme deficiencies. The National Urea Cycle Disorders Foundation agreed. When Batshaw turned up at their 1994 annual meeting asking for volunteers, so many mothers offered to be screened for the OTC gene that it took him four hours to draw all the blood.

———

BY THE TIME Mark Batshaw and Jim Wilson submitted their experiment to the Recombinant DNA Advisory Committee for approval, the panel was in danger of being disbanded. Varmus, the NIH director, who won the Nobel Prize for his discovery of a family of cancer-causing genes, had made no secret of his distaste for the conduct of gene-therapy researchers. He thought the science was too shoddy to push forward with human testing, and it bothered him that so few experiments were focusing on genetic diseases. It irked him to have to sign off on protocols the RAC approved, and it irked him even more to see biotech companies touting those approvals, like some kind of NIH imprimatur, in the business pages of the papers. "Some days," says Dr. Nelson Wivel, the committee's former executive director, who now works for Wilson at Penn, "it felt as though the RAC was helping the biotech industry raise money. Dr. Varmus hated that."

At the same time, the pharmaceutical industry and AIDS activists were complaining that the RAC was redundant: the FDA already reviewed gene-therapy proposals. So in mid-1995, after seeking the advice of an expert panel, Varmus reorganized the RAC, slashing its membership from twenty-five to fifteen and stripping it of its approval authority—a decision that, some say, has enabled gene-therapy researchers to ignore the panel and keep information about safety to themselves. "The RAC," complains Dr. Robert Erickson, a University of Arizona medical geneticist who served on the panel, "became a debating society."

The Batshaw-Wilson protocol was among the last the committee would ever approve. The plan was for eighteen adults (nineteen eventually signed up, including Tish Simon, but the last patient was never treated, because of Jesse's death) to receive an infusion of the OTC gene, tucked inside an adenovirus vector, through a catheter in the hepatic artery, which leads to the liver. The goal was to find what Wilson calls "the maximum tolerated dose," one high enough to get the gene to work, but low enough to spare patients serious side effects. Subjects would be split into six groups of three, with each group receiving a slightly higher dose than the last. This is standard fare in safety testing. "You go up in small-enough increments," Wilson explains, "that you can pull the plug on the thing before people get hurt."

The experiment stood in stark contrast to others that had earned Varmus's scorn. It was paid for by NIH, which meant it had withstood the rigors of scientific peer review. It was aimed at a rare genetic disease, not cancer or AIDS. It was supported by plenty of animal research: Wilson and his team had performed more than twenty mouse experiments to test efficacy and a dozen safety studies on mice, rhesus monkeys and baboons. Still, it made Erickson, one of two scientists assigned by the RAC to review the experiment, uneasy.

He was troubled by data showing that three monkeys had died of a blood-clotting disorder and severe liver inflammation when they received an earlier, stronger version of the adenovirus vector at a dose twenty times the highest dose planned for the study. No one had injected adenovirus directly into the bloodstream before, either via the liver or otherwise, and the scientists admitted that it was difficult to tell precisely how people would respond. They planned to confine the infusion to the right lobe of the liver, so that if damage occurred it would be contained there, sparing the left lobe. And they outlined the major risks: bleeding, from either the gene-therapy site or a subsequent liver biopsy, which would require surgery; or serious liver inflammation, which could require an organ transplant and might lead to death.

Both Erickson and the other scientific reviewer thought the experiment was too risky to test on asymptomatic volunteers and recommended rejection. But in the end, Batshaw and Wilson prevailed. They offered up Caplan's argument that testing on babies was inappropriate. And they agreed to inject the vector into the bloodstream, as opposed to putting it directly into the liver. That decision, however, was later reversed by the FDA, which insisted that because the adenovirus would travel through the blood and wind up in the liver anyway, the original plan was safer.

The RAC, in such disarray from Varmus's reorganization that it did not meet again for another year, was never informed of the change.

JESSE GELSINGER WAS SEVENTEEN when his pediatric geneticist, Dr. Randy Heidenreich, first told him about the Penn proposal. He wanted to sign up right away. But he had to wait until he was eighteen.

Paul Gelsinger was also enthusiastic. A trim forty-seven-year-old with intense blue eyes, Gelsinger, who makes his living as a handyman, gained custody of his four children nine years ago, when he divorced their mother, who suffers from manic depression. He had been having some difficulty with Jesse then; the boy was in the midst of an adolescent rebellion and was refusing to take his medicine. "I said: 'Wow, Jess, they're working on your disorder. Maybe they'll come up with a cure.' "

Jesse's was not a typical case of OTC deficiency: his mutation appears to have occurred spontaneously in the womb. His disease having been diagnosed when he was two, Jesse was what scientists call a mosaic—a small portion of his cells produced the missing enzyme. When he watched what he ate and took his medicine, he was fine. But one day last December, Paul Gelsinger arrived home to find his son curled up on the couch. He had been vomiting uncontrollably, a sign, Paul knew, that Jesse's ammonia was rising. Jesse landed in the hospital, comatose and on life support. When he recovered, he never missed another pill.

On June 18, the day Jesse turned eighteen, the Gelsingers—Paul, Mickie and the children—flew to Philadelphia to see Paul's family. They played tourists, visiting the Liberty Bell and the Rocky statue, where Jesse was photographed, fists raised, a picture that would circulate in the newspapers after his death. On the twenty-second, they went to the University of Pennsylvania, where they met Raper, the surgeon, who explained the experiment and did blood and liver-function tests to see if Jesse was eligible. He was, and his treatment was scheduled for the fall. Jesse would be the youngest patient enrolled.

On September 9, Jesse returned to Philadelphia, this time alone. He took one duffel bag full of clothes and another full of wrestling videos. Paul Gelsinger planned to fly in a week later for the liver biopsy, which he considered the trial's most serious risk.

The treatment began on Monday, September 13. Jesse would receive the highest dose. Seventeen patients had already been treated, including one woman who had been given the same dose that Jesse would get, albeit from a different lot, and had done "quite well," Raper says. That morning, Jesse was taken to the interventional-radiology suite, where he was sedated and strapped to a table while a team of radiologists threaded two catheters into his groin. At 10:30 A.M., Raper drew 30 milliliters of the vector and injected it slowly. At half past noon, he was done.

That night, Jesse was sick to his stomach and spiked a fever, 104.5 degrees. Raper was not particularly surprised: other patients had experienced the same reaction. Paul Gelsinger called; he and Jesse talked briefly, exchanging I love yous. Those were the last words they ever spoke.

Early Tuesday morning a nurse called Raper at home; Jesse seemed disoriented. When Raper got to the hospital, about 6:15 A.M., he noticed that the whites of Jesse's eyes were yellow. That meant jaundice, not a good sign. "It was not something we had seen before," Raper says. A test confirmed that Jesse's bilirubin, a breakdown product of red blood cells, was four times the normal level. Raper called Gelsinger, and Batshaw in Washington, who said he would get on a train and be there in two hours.

Both doctors knew that the high bilirubin meant one of two things: either Jesse's liver was failing or he was suffering a clotting disorder in which his red blood cells were breaking down faster than the liver could metabolize them.

This was same disorder the scientists had seen in the monkeys that had been given the stronger vector. The condition is life-threatening for anyone, but particularly dangerous for someone with Jesse's disease, because red blood cells liberate protein when they break down.

By midafternoon Tuesday, a little more than twenty-four hours after the injection, the clotting disorder had pushed Jesse into a coma. By 11:30 P.M., his ammonia level was 393 micromoles per liter of blood. Normal is 35. The doctors began dialysis.

Paul Gelsinger had booked a red-eye flight. When he arrived in the surgical intensive care unit at 8:00 Wednesday morning, Raper and Batshaw told him that dialysis had brought Jesse's ammonia level down to 72 but that other complications were developing. He was hyperventilating, which would increase the level of ammonia in his brain. They wanted to paralyze his muscles and induce a deeper coma, so that a ventilator could breathe for him. Gelsinger gave consent. Then he put on scrubs, gloves and a mask and went in to see his son.

By Wednesday afternoon, Jesse seemed to be stabilizing. Batshaw went back to Washington. Paul felt comfortable enough to meet his brother for dinner. But later that night Jesse worsened again. His lungs grew stiff; the doctors were giving him 100 percent oxygen, but not enough of it was getting to his bloodstream. They consulted a liver-transplant team and learned that Jesse was not a good candidate. Raper was beside himself. He consulted with Batshaw and Wilson, and they decided to take an extraordinary step, a procedure known as ECMO, for extracorporeal membrane oxygenation, essentially an external lung that filters the blood, removing carbon dioxide and adding oxygen. It had been tried on only 1,000 people before, Raper says. Only half had survived.

"If we could just buy his lungs a day or two," Raper said later, they thought "maybe he would go ahead and heal up."

The next day, Thursday, September 16, Hurricane Floyd slammed into the East Coast. Mickie Gelsinger flew in from Tucson just before the airport closed. (Pattie Gelsinger, Jesse's mother, was being treated in a psychiatric facility and was unable to leave.) Batshaw spent the day trapped outside Baltimore on an Amtrak train. He ran down his cell phone calling Raper; when it went dead, he persuaded another passenger to lend him his. The ECMO, Raper reported, appeared to be working. But then another problem cropped up: Jesse's kidneys stopped making urine. "He was sliding into multiple-organ-system failure," Raper says.

That night, at his hotel, Paul Gelsinger couldn't sleep. He left his wife a note and walked the half mile to the Penn medical center to see Jesse. The boy was bloated beyond recognition; even his ears were swollen shut. Gelsinger noticed blood in Jesse's urine, an indication, he knew, that the kidneys were shutting down. How can anybody, he thought, survive this?

On the morning of Friday the seventeenth, a test showed that Jesse was brain dead. Paul Gelsinger didn't need to be told: "I knew it already." He called for a chaplain to hold a bedside service, with prayers for the removal of life support.

The room was crowded with equipment and people: seven of Paul's fifteen siblings came in, plus an array of doctors and nurses. Raper and Batshaw, shellshocked and exhausted, stood in the back. The chaplain anointed Jesse's forehead with oil, then read the Lord's Prayer. The doctors fought back tears.

When the intensive-care specialist flipped two toggle switches, one to turn off the ventilator and the other to turn off the ECMO machine, Raper stepped forward. He checked the heart-rate monitor, watched the line go flat and noted the time: 2:30 P.M. He put his stethoscope to Jesse's chest, more out of habit than necessity, and pronounced the death official. "Good-bye, Jesse," he said. "We'll figure this out."

WILSON REPORTED THE DEATH immediately, drawing praise from government officials but criticism from Arthur Caplan, who says they should have made the news public, in a news conference. In the weeks since, the Penn team has put every detail of Jesse's treatment under a microscope. It has rechecked the vector to make certain it was not tainted, tested the same lot on monkeys, reexamined lab and autopsy findings. Wilson's biggest fear was that Jesse died as a result of human error, but so far there has been no evidence of that. "That's what's so frightening," French Anderson says. "If they made a mistake, you would feel a little safer."

The death has rattled the three doctors in various ways. Wilson has asked himself over and over again whether he should have done anything differently. "At this point, I say no, but I'm continuing to re-evaluate constantly." He has been besieged by worry, about the morale of his staff, about whether his institute's financial sponsors would pull out, about whether patients would continue to volunteer, about whether he would lose his bravado—the death knell for a scientist on the cutting edge. "My concern," he confessed, over dinner one night in Philadelphia, "is, I'm going to get timid, that I'll get risk-averse."

Raper has thrown himself into his work, trying to live up to his promise to "figure this out." There are a number of possible explanations, he says: the vector may have reacted badly with Jesse's medication; Jesse's status as a mosaic may have played a role; or perhaps the early testing in monkeys, which showed that the stronger vector had deleterious side effects, was more of a harbinger of danger than the doctors realized. An answer may take months, but he is determined to find one; only by understanding what happened to Jesse, and how to prevent it in others, can the research continue. "That," Raper says, "would be the best tribute to Jesse."

Of the three, Batshaw seems to have taken it the hardest. He is not a particularly religious man, but a few days after Jesse died he went to synagogue to say Kaddish, the Jewish mourner's prayer. He struggles with the idea of personal responsibility. He has cradled many a dying child in his career, but never before, he says, has a patient been made worse by his care. "What is the Hippocratic oath?" Batshaw asks rhetorically, looking into the distance as his fingers drum the tabletop. He pauses, as if to steel himself, and says, "I did harm."

Paul Gelsinger does not hold the doctors responsible, although he is acutely interested in knowing what other scientists knew about adenovirus before Jesse died. He has experienced a deep spiritual awakening since losing his son; in dying, he says, Jesse taught him how to live. He speaks frequently of God, and of "purity of intent," which is his way of saying that Jesse demonstrated an altruism the rest of us might do well to emulate. "I hope," he said on the mountaintop that Sunday afternoon, "that I can die as well as my son has died."

DEBORAH M. GORDON

Close Encounters

FROM *THE SCIENCES*

*The behavior of ants is one of nature's most fascinating and puzzling phe-
nomena. How do individual ants come together to form large, organiza-
tionally complex colonies? Stanford University biologist Deborah Gordon,
who for several years has closely observed ants under the blazing sun of the
Arizona desert, has made waves with some surprising findings that challenge previously held theories of how ants choose tasks and divide labor.*

Every summer for the past seventeen years I have studied the harvester
ants in a small patch of the Arizona desert. The site is at the side of a
rough paved road that runs through a flat valley between the Chiricahua
and Peloncillo mountains at the state line of Arizona and New Mexico. An
enormous sky surrounds an endless reach of land. The Chiricahuas, to the
west, are so close you can see the patches of rock change color during the day.
The Peloncillos, to the east and north, form a jagged outline in the distance. To
the south the desert stretches eighty miles to the Mexican border.

I stay at the Southwestern Research Station up in the Chiricahua Mountains. The station belongs to the American Museum of Natural History in New
York City. At the peak of the summer season fifty people might be staying
there, mostly undergraduates who come to work either for the station or as research assistants for people like me. We get up at 4:30, when it is still dark, and
meet in the dining room. Then there is some antlike milling around while peo-

ple make their peanut-butter-and-jelly sandwiches to get them through the morning, collect their water bottles and clipboards, and pile into the van. It's about a twenty-minute drive down into the desert.

You can recognize an ant field-worker by looking at her ankles: We wear our socks over the cuffs of our pants. Experience with stinging ants teaches that you can see them on your hands and feel them go down the back of your neck, but they can get up inside your pants faster than you can brush them off. In the desert, however, keeping off the sun is more difficult than keeping off the ants. Some people wear shorts, at least until they have been stung. Over the years I have evolved a costume that includes a long-sleeved shirt, a cap with a kind of curtain around its lower edge, and the largest sunglasses I can find. I look rather like an insect myself.

I STUDY ANTS because I am interested in linking levels of organization. The living world is structured in layers, from molecules to cells to organs to individuals to populations to ecosystems. A fundamental question in biology is how events at those different levels are related. A thought happens when electrical impulses move around the tangle of neurons in a brain, but a thought is something more than, and something other than, neurons. I set out to do research on the connections between the tiers of natural organization, and I chose to work with ants because they invite the observer to notice three distinct levels: ants, colonies and populations.

All ants live in colonies consisting of sterile workers and one or more queens. The queens produce the workers and the next generation of reproductives: new queens and males. The new reproductives fly off to mate and then begin new colonies. The population is made up of all of the colonies whose reproductives could mate with one another.

In studying ant behavior I try to figure out how each ant decides what to do, and how this adds up to the achievements of colonies. Because ants are separate beings that move around freely, they attract attention as individuals. But nothing ants do makes sense except in the context of the colony. Zoom in, you see ants—zoom out, you see a colony. Ants and colonies are both there in front of you, all the time.

Ant-colony life has evolved over the past 100 million years from the less social lives of the ants' ancestors, the wasps. The evolution of colony behavior hinges on how well a colony functions, relative to the other colonies in its population. If a colony that acts a certain way tends to produce more offspring, and variation in colony behavior is heritable, then over many generations the more successful behavioral types will be better represented in the population than less successful ones. In fact, biologists know nothing about the inheritance of ant-colony behavior, and little about the past conditions in which the behavior we now see once evolved.

As a kind of shortcut over those chasms of ignorance, one can pretend that success in the ant world is the production of many new colonies. If natural selection is acting on ant behavior now, one can see it only by asking ecological questions: Why might some colonies reproduce more than others? How does a colony get resources and how does it use them? Where does it live and what does it eat? What eats it and with what does it share food and space?

My long-term study has made it possible to link the behavior of ants within colonies to the ecology of populations. I ask how the behavior of an ant fits into the life of its colony, and how in turn the relations of neighboring colonies may shape, over evolutionary time, the behavior of ants within colonies. Because I have kept track of about 300 colonies for many years, I have followed colonies through their life cycles. New colonies are born, struggle to occupy a foraging area, grow larger, start to reproduce and then settle in among their lifelong neighbors.

I discovered that how colonies behave and how they relate to their neighbors both change as a colony grows older and larger. Neighboring colonies search the same places for food, so pressure from neighbors can starve out a small, new colony or keep an older one from reproducing. Colonies live for fifteen to twenty years, but each ant lives only a year. When an older colony acts differently from a younger one, it is not following the advice of older, more experienced ants. This leads me to ask why, if the ants in young colonies behave just like ants in old ones, the colonies seem to become more staid and prudent as they get older and larger.

It is difficult to think about how an ant colony works. Not only is an ant colony's behavior complex, woven from zillions of ant acts, but all those tiny events add up to something different from any society we know. Stories about totalitarian societies, inexorable armies and voracious monsters are often told as stories about ants. But ants have no dictators, no generals, no evil masterminds. In fact, there are no leaders at all. The relations between neighboring colonies are as intriguing as the inner workings of a single ant society. Ecological accounts of animals' lives often portray a kind of suburban nightmare, each individual or family struggling furiously to accumulate more and better resources than its neighbors—with the richest one the evolutionary winner.

But ants have no property, no bank accounts, no picket fences. In a harvester ant neighborhood, there are no borders. This is the puzzle. If ants don't work as a miniature human society, how do a group of rather inept little creatures create a colony that gets things done?

THE TWENTY-FIVE ACRES that I have inspected inch by inch for the past seventeen summers is part of a 7,000-acre cattle ranch. On the human scale, the site looks like a plain of chaparral scrub with mountains on either side. On the ant-colony scale, the site looks like a bumpy, sandy surface with lots of gray, orange and pink boulders and the occasional bush or plant. By the time it gets warm, about 7:30 in the morning, the ground quietly teems with harvester ants. They are called harvester ants because they eat mostly seeds, which they store inside their nests. They will happily take termites as well when they can find them.

There is a nest of the red harvester ant (*Pogonomyrmex barbatus*) every few yards. Young harvester-ant colonies often start out with nests under bushes; eventually they destroy the bush roots as the nest expands underground, creating an open space by the time the colony is four or five years old. An old, established colony may have a flat disk or mound a meter wide, covered with tiny pebbles, with one or two entrances in the middle. Most colonies stay in the same nest all their lives.

P. barbatus ants are large—about a centimeter long—and brown. They have big heads that make them appear to jostle along as they walk. The back end does the propulsion and at every step pushes the heavy head forward with a little bump. The ants are sedate, even plodding. They spill out of the nest in the morning like elderly tourists pouring out by the busload. There are many species of *Pogonomyrmex* in the deserts of the southwestern United States and northern Mexico, and they are all known for their powerful sting. Perhaps because their potent venom makes aggression unnecessary, they rarely attack, and they tend to avoid confrontation.

As many as fifty other ant species may live on the site. But my view of them is Pogo-centric, in that I think of them as belonging to three categories: other harvester-ant species in the genus *Pogonomyrmex,* other big ants, and little ants. The differences in size may not matter as much to the ants as they do to me; my categories reflect how well I can see them. Many species of tiny ants have been studied in the laboratory under a dissecting microscope, but I prefer to watch animals I can see without having to fiddle with any equipment.

A HARVESTER-ANT COLONY carries out many tasks: it must collect and distribute food, build a nest, and care for its eggs, larvae and pupae. It lives in a changing world to which it must respond. When there is a windfall of food, more foragers are needed. When the nest is damaged, extra effort is required for quick repairs.

Task allocation is a process that operates without any central or hierarchical control to direct individual ants into particular tasks. The queen exercises no authority; she merely lays eggs and is fed and cared for by the workers. No ant can assess the global needs of the colony, or count how many workers are engaged in each task and decide how many should be allocated differently. Yet there is abundant evidence throughout physics, the social sciences and biology, that simple behavior by individuals can lead to predictable patterns in the behavior of groups. It should be possible to explain task allocation in a similar way, as the consequence of simple decisions by individuals.

I began to study the dynamics of task allocation by asking how tasks are related. Does a change in one task lead to a change in another? To answer that question, I conducted a series of perturbation experiments. In one such experiment, I put small plastic barriers down on the foraging trails, at a place far enough from the nest that only foragers would go there. The ants could not go over the barriers, but they could go around them or through them via a few holes in the bottom. Thus the barriers were not impermeable, but they did create an eddy in the flow of foragers, slowing down foraging traffic on the nest side of the barrier.

Colonies might have reacted to the barriers by sending out even more foragers to compensate for the obstacle on the trail, but they did not. The decrease in foraging in the presence of barriers probably mimics the response of the colony to any event that slows down food intake. Foraging is a costly activity; ants use up water and energy when they move around the hot desert, searching for seeds. A decline in the rate at which foragers bring in food may signal a scarcity of food, when it is not worthwhile to forage.

A second perturbation experiment directly involved only nest-maintenance workers. Early in the morning, when those workers were first active, I put small piles of toothpicks next to the nest entrance. Within about forty minutes nest-maintenance workers carried all the toothpicks to the edge of the nest mound and left them there. Carrying toothpicks required extra nest-maintenance workers, and so the numbers of those workers increased.

The experiments showed that foraging and nest-maintenance are strongly related. When the number of foragers was experimentally decreased by the barriers placed on the foraging trails, the number of nest-maintenance workers increased. Even though the nest-maintenance workers were not directly affected by the barriers, some interaction between nest-maintenance workers and foragers must regulate the number of nest-maintenance workers. When the number of nest-maintenance workers was experimentally increased, by the introduction of toothpicks, the number of foragers decreased. Again, though foragers were not directly affected by the presence of toothpicks, they responded to changes in the number of workers engaged in nest maintenance.

To find out whether ants switch tasks, I did the perturbation experiments again, with ants that had been marked with drops of model-airplane paint. Before marking the ants, my coworkers and I first cooled them down so that we could handle them without being stung, and so that they would not wave their antennae around and get them stuck in the wet paint. Making ants too cold to move was not easy in the desert, but eventually I found the ideal piece of equipment: an ice-cream maker that stays cold for a long time after being frozen. We used paints of several colors, each color corresponding to one task. We collected ants as they foraged, patrolled, did nest-maintenance work or worked on their midden, or refuse pile. Then we marked them, let the paint dry and released them. The next day we used some of the colonies with marked ants to do perturbation experiments and left others undisturbed as controls.

In colonies that were not subjected to any experiments, few ants switched tasks. Foragers marked one day tended to forage the next, and so on for patrollers, nest-maintenance workers and midden workers. But when my experiments created a need for extra workers, the ants switched tasks.

THE FINDING THAT ANTS switch tasks stood in contrast to earlier work, which had postulated that an individual ant would carry out a single task throughout its life. An individual's "caste" was its inherent tendency to do a particular task. The assumption that an individual's task is fixed reinforced a view of individuals as independent of one another. The problem posed by that caste-oriented research program was to explain how natural selection acts on the colonies of a given species to produce optimal distributions of workers in each caste. The number of workers performing a task was considered to be a simple function of the number of specialized individuals in the corresponding caste. For example, the amount of foraging a colony does would correspond to the number of foragers it contains.

But the number of ant workers that perform a task is not a simple function of the number of ants of that task group. An ant may specialize for a short time, but it often does several tasks, not just one. The number that perform a given task right now depends on rules that determine which ants will switch to perform the task in the current situation. How much a colony forages may depend in part on how many ants are available to work as foragers, but it also depends on the number engaged in nest-maintenance work, the number patrolling and so on. Somehow, the behavior of ants in other task groups influences the probability that an ant will be active and the probability that it will perform a particular task.

One way to investigate task allocation is to model it, then generate hypotheses from the models that can be tested empirically. The models my colleagues and I have developed so far, including neural networks as well as more deterministic models, show how task allocation could work if an ant's behavior depends on simple cues from its environment and from its interactions with other ants. Whether harvester ants actually behave as the models predict is an empirical question.

Although workers engaged in one task are unlikely to encounter cues that indicate the status of other tasks, they are very likely to encounter workers from other task groups. Colonies are active outside the nest for six to seven hours each day, and all exterior workers frequently go in and out of the nest. A trip by a forager can take about half an hour, and nest-maintenance workers, patrollers and midden workers make even shorter trips outside the nest. The exterior workers rarely travel deep inside the nest, so workers of all task groups mix in the upper chambers of the nest, directly inside the entrance.

THE NUMBER OF ANTS active outside the nest at any given time, therefore, influences the number of interactions experienced by all exterior workers: both the active ones, when they enter and leave the nest, and the currently inactive ones, which are simply waiting inside.

Suppose the probability that an ant performs a task depends on the recent history of its encounters. The ant would have to distinguish between encounters with ants of different task groups, and we know that task groups differ in odor. The probability that the ant will perform the task depends on a rule such as: "If I meet five successful foragers, go out to forage; if I meet fewer than five foragers, remain inactive." The decrease in the number of ants leaving the nest to forage when I took the food away from returning foragers suggests that some such rule is at work: ants that met too few successful foragers were less likely to leave the nest to forage.

My coworkers and I have studied colonies in the laboratory as well as in the field, to find out how an ant's recent interactions influence its task. We mark the ant, follow it and record all its antennal contacts with other ants. Then we analyze our data to see if there is any relation between the ant's contacts and what that ant does next. We found that the more contact an ant has with midden workers, the more likely it is to do midden work. The rate as well as the number of contacts is important.

Although colonies must respond to changing conditions, task allocation does not have to be perfect. It is not like clockwork, or an army, each unit snapping into place so the whole system ticks on without a hitch. There must be enough ants to collect food often enough for the colony to survive and grow, but if the colony does not get enough food today, perhaps it will tomorrow. The allocation process results in more or less the right number of ants engaged in the appropriate task, often enough for the colony to carry on.

The most difficult thing to grasp about task allocation is that it is not a deterministic process even at the individual level. An ant does not respond the same way every time to the same stimulus; nor do colonies. Some events influence the probabilities that certain ants will carry out certain tasks, and this regularity leads to predictable tendencies rather than perfectly deterministic outcomes. The ant is jostled along in a stream of events that send it sometimes into one task, sometimes another, like a twig in a turbulent river.

It is tempting to speculate about the generality of interaction patterns as a source of information in natural systems. What I like about the idea that an ant's task decision is based on its interaction rate is that the pattern of interaction, not a signal in the interaction itself, produces the effect. Ants do not tell one another what to do by transferring messages. The signal is not in the contact, or in the chemical information exchanged in the contact. The signal is in the pattern of contact. Such a process might operate in brains, immune systems, or anyplace where the rate of flow of a certain kind of unit, or the activity level of a certain kind of unit, is related to the need for a change in the rate of flow.

PERHAPS ANTS HAVE SOMETHING general to teach us, at least by analogy, about how nature works. Any system that is made up of units lacking identity or agency, and whose behavior arises from the interactions between those components, has something in common with ant colonies. It may be that the same kinds of relations that link ants and colonies allow neurons to produce the behavior of brains; a host of different cells to produce immune responses; and a few dividing cells to produce a developed embryo.

Ants do not offer moral instruction, but they show how simple parts make complex living systems, and how those systems connect to the outside world. Looking at ants in colonies while looking at colonies in populations, we see how the layers of a natural system fit together.

Francis Halzen

Antarctic Dreams

FROM *THE SCIENCES*

*The subatomic particle known as the neutrino is nearly impossible to de-
tect. Having almost no mass, it passes through most matter without touching it. In order to capture the rare collision between a neutrino and another
particle, physicists have built enormous detectors, whose scales seem comi-
cally disproportionate to the quarry they are designed to hunt. Francis
Halzen, a physicist at the University of Wisconsin, is the architect of a neutrino detector that acts as a telescope, but, just to make things even more
complicated, it had to be built in the remote reaches of the South Pole.
Halzen's report, like Deborah Gordon's, describes the hardships and uncertainties scientists encounter in seeking valuable data.*

Big science, as often as not, hinges on small moments. Once the grants
have been secured and the politics navigated, the ground broken and
the visionary promises made, when banks of computers flicker to life
and fingers curl above keyboards, ready to flash the first discoveries via E-mail,
a decade's work can still come crashing down in the final hour—a castle
erected in the thin air of theory, too weak to withstand true gravity.

Last August, when Dennis S. Peacock came to visit my office at the University of Wisconsin–Madison, he must have had something like that in the
back of his mind. As the head of the Antarctic sciences section for the National
Science Foundation (NSF), Peacock had helped funnel some $10 million into

building the world's largest neutrino telescope deep in the Antarctic ice. Would we have anything to show for the investment? Or would our project go the way of its predecessor, partly built in seawater off the coast of Hawai'i and then abandoned after nearly twenty years of work?

Our telescope—also known as the Antarctic Muon and Neutrino Detector Array, or AMANDA—had been detecting neutrinos for some time, but we had yet to finish analyzing our first data. Or so I thought. Instead, when Peacock and I arrived at the desk of my graduate student, Rellen R. Hardtke, she had a surprise waiting for me. Two of her colleagues had spent the night finishing the analysis, she explained with a smile. Then she calmly called up an image on her computer screen: the first precision map of a high-energy neutrino event ever recorded.

Hardtke's screen showed a faint blue line streaking diagonally across columns of black dots. Most of the dots, each of which represented a photomultiplier sunk in the ice, were small and black. But a few, clustered along the line, were blue or green or red, and two, near the beginning of the line, were bright orange and very large. At five in the morning on October 12, 1997, the diagram told us, a neutrino—one of nature's smallest and most elusive elementary particles—had entered the earth in the middle of the Pacific Ocean, between Midway Island and the Aleutians, hurtled straight through the planet, and collided head-on with a proton on the underside of the Antarctic ice. Two kilometers beneath the surface, our grid of photomultipliers had picked up a subatomic spark from that collision as it flew upward through the ice and flared past them for about a microsecond.

The end result, abstracted on a computer screen, would have seemed unremarkable to most people. But to us it was the sole trace of a particle that had traversed vast distances to reach us, perhaps a remnant of one of the most spectacular events in the universe. More to the point, it was the first concrete proof that AMANDA worked—that it could help map the distant depths of space, thereby perhaps resolving some of the most heated controversies in physics. Later, when I E-mailed the diagram to our collaborators, one of them wrote back: "*This* is why I've spent five years of my life on this project."

Neutrinos, they are very small," John Updike wrote, famously. "They have no charge and have no mass / And do not interact at all." As it turns out, Updike was wrong on two counts, but he got the spirit right, anyway. Neutrinos are so small and slippery that they pass through the earth (and stars and cities and most everything else) like a bullet through a rainstorm. Unfazed by magnetic fields or the strong nuclear force, they have to make a direct hit on a proton to be stopped at all—a highly unlikely event. At the same time, neutrinos are about as plentiful in the universe as the photons that constitute light: 3×10^{16} of them pass through our bodies every second.

That combination of factors makes neutrinos the best focus for deep-space astronomy. Most photons cannot reach us from the most distant points in the universe (the most energetic ones cannot even make it from the edge of the galaxy: they crash into microwave background radiation along the way). And though radio waves routinely travel as far as neutrinos, they are emitted even by rather mundane astronomical objects—the moon, for instance. Neutrinos, on the other hand, not only travel long distances, they are easy to categorize: low-energy neutrinos are generated by the sun, by cosmic-ray collisions in the upper atmosphere, and by other nearby phenomena; high-energy neutrinos only reach the earth from distant, supremely violent events—gamma-ray bursts, for instance, or black holes at the center of new galaxies. By focusing on high-energy neutrinos alone, a telescope can naturally filter out all but the most interesting things in the sky.

But there is no free lunch. If neutrinos can fly through planets without stopping, they hardly brake for your average telescope mirror. In fact, neutrinos are so hard to detect that for decades they existed only in theory. The Swiss theoretical physicist Wolfgang Pauli "invented" them in 1930 to balance out the energy apparently lost when radioactive matter decays. ("I have done a terrible thing," he told the German astronomer Walter Baade. "I have postulated a particle that cannot be detected.") It was not until twenty-six years later, when the physicists Frederick Reines and Clyde L. Cowan built a neutrino detector near the Savannah River nuclear plant in South Carolina, that the existence of the particle was confirmed.

IN THE PAST FOUR DECADES neutrino detectors of increasing scale and sensitivity have cropped up in odd spots around the globe: in an iron mine in Minnesota; underneath a mountain in Italy's Apennine range; in an abandoned railway tunnel on the outskirts of Osaka, Japan. None of them are really telescopes, however: rather than tracking high-energy neutrinos to map deep space, as AMANDA does, they simply detect low-energy neutrinos from the sun.

One might think that one or two solar neutrino detectors would be enough. But a single mystery has continued to tantalize investigators. According to standard astrophysical theory, nuclear fusion inside the sun ought to spawn a stable number of neutrinos, and so physicists ought to detect a predictable number of them. Instead, month after month, in detector after detector, no more than half of the expected neutrinos are counted. Are the detectors faulty, or does solar astrophysics need some revision? Neither one, most physicists now say. The flux of solar neutrinos is too weak because the particles transform themselves en route to the earth. What were once electron neutrinos become their particulate cousins: muon neutrinos and tau neutrinos, which most neutrino detectors cannot detect.

The fact that neutrinos come in three "flavors" is old and undisputed. But the idea that they oscillate between those three flavors entails a fundamental rethinking of their nature—and perhaps of the universe itself. According to the standard model of elementary particle physics, neutrinos have no mass at all. Yet according to quantum theory, only particles with mass can oscillate between one flavor and another. By recent estimates, there are about 100 million times as many neutrinos in the universe as there are protons and neutrons combined. Even if the mass of each neutrino is no more than a tenth of an electron volt, their collective mass would be as great as that of all the visible matter in the universe.

"Neutrino oscillations have been discovered at least four times and undiscovered at least twice," notes the particle physicist Donald H. Perkins of the University of Oxford. Last June, however, physicists in Takayama, Japan, announced results that have silenced most of the remaining doubters. Their Super-Kamiokande detector, built in a working zinc mine a kilometer underground, incorporates more than 13,000 photomultipliers to survey 50,000 tons of water for telltale flashes of light. (A photomultiplier looks like a lightbulb and works like a lightbulb in reverse: light goes in and electricity comes out. But what a lightbulb! The photomultipliers in Super-K amplify signals by a factor of 100 million.)

In principle, Super-K ought to detect equal numbers of low-energy neutrinos radiating in from all sides from the collisions of cosmic rays with the atmosphere. But in the past three years, the detector found that fewer neutrinos were coming in from the farside of the earth than from the nearside. The only explanation for the discrepancy, the physicists concluded, was that some muon neutrinos, which Super-K can detect, must have changed into undetectable tau neutrinos as they passed through the earth.

The Super-K data are so precise, so elegantly conceived, that they seem to prove once and for all that neutrinos have mass—thereby punching a hole in one of the standard model's tires. Whether fixing that hole will require a simple patch or a complete overhaul remains to be seen: Super-K can only measure the *difference* in mass between two neutrinos—a quantity somewhere between 0.1 and 0.01 electron volt. Based on that figure and other data, however, many physicists now estimate that the three types of neutrinos have a combined mass of around 0.1 electron volt.

What Super-K does not do—what no working detector has ever done—is pay attention to neutrinos from beyond our galaxy. Enter AMANDA. Because high-energy neutrinos are 10^{12} times easier to detect than solar neutrinos, our telescope can afford to trade sensitivity for size. The same volume of water surveyed by 13,000 photomultipliers in Super-K is surveyed by only around ten in AMANDA; our photomultipliers are less than half as large; and ours do not detect solar neutrinos at all. But AMANDA, when complete, will watch over thousands of times more water than Super-K. As a result, it will track neutrinos across as far as a kilometer, whereas Super-K can track them across no more than fifty meters.

When a high-energy neutrino collides with a proton in the ice, it creates a muon—a particle closely related to the electron but more than 200 times more massive—that continues along the neutrino's upward path, streaming photons along its sides like a bottle rocket. The result is a hurtling cone of blue Cherenkov light—light of the same kind emitted by nuclear reactors. From the timing of the cone's reception by our grid of photomultipliers, the muon's direction can be reconstructed and the direction of the incoming neutrino inferred.

Why are we looking for neutrinos that have passed through the earth? Muons generated by cosmic rays also bombard the earth constantly from every direction, and some of them can travel through miles of ice before petering out. But no muon can cross the entire planet—as much as 8,000 miles of dense iron, magma and rock. Hence by pointing its photomultipliers downward, AMANDA uses the earth itself to screen out all upward-traveling muons except the ones thought to come from high-energy neutrinos.

As early as the 1960s, physicists had dreamed that radio antennas, operating near gigahertz frequencies, might listen in on the electric charges sparked by neutrinos crashing into ice. Thirty-five years later, however, working through the theory behind that idea, my colleagues and I showed that the radio signal created by the neutrinos was too weak to be of any use. It was then that I hit upon the obvious alternative: Why not try to detect the flash from a neutrino collision, rather than its noise?

I suspect that others must have contemplated the same idea and given up on it. Had I not been completely ignorant of what was then known about the optical properties of natural ice, I would probably have done the same. Instead, I sent off a flurry of E-mail messages to my friend John G. Learned, then the spokesperson for the Deep Underwater Muon and Neutrino Detector (DUMAND). Like AMANDA after it, DUMAND was designed to detect high-energy muons—though in ocean water off the coast of Hawai'i rather than in Antarctic ice. But though Learned's group had already deployed a test string of photomultipliers with some success, the project was eventually abandoned. (The DUMAND concept lives on, albeit at a smaller scale, in a detector now operating in the depths of Lake Baikal in Russia.)

Learned immediately appreciated the advantages of an Antarctic neutrino telescope. For starters, sinking the photomultipliers into ice would enable investigators to walk around on top of the experiment, as well as to keep all the fragile electronics at the surface. As a result, the neutrino signals could be identified with off-the-shelf electronics. Better yet, the ice would be geologically stable (Antarctica almost never has earthquakes) and completely dark. DUMAND did not have it so easy. Although the water off the Hawaiian coast is exceptionally clear and deep, Learned's group had to contend with waves, storms, background light from bioluminescent organisms and the radioactive decay of sea salt. Most important, NSF was already operating a research station at the geographical South Pole, with an infrastructure to rival that of a national particle physics laboratory. AMANDA, in other words, would be much cheaper and easier to build than DUMAND.

Still, convincing donors of the soundness of our idea was no simple matter: I was only a theorist, after all, with no experience building anything, and my collaborators, at least in the beginning, were very talented but very junior physicists at the University of California, Berkeley. Nevertheless, NSF was willing to give us the benefit of the doubt, and within a few years we had joined forces with eight other universities and three research laboratories in Belgium, Germany, Sweden and the United States. In 1990, as proof of principle, one of our teams sank a 200-meter-long strand of three photomultipliers into the packed snow of Greenland. The toy experiment detected muons. Then, in the Antarctic summer of 1992, our work began in earnest.

———

As I write, it is ten degrees below zero Fahrenheit outside my office in Madison, and I am dressed just warmly enough to be slightly cold. In most places, I think, people have made an art of underdressing in winter. But not in Antarctica. Faced with temperatures that regularly dip to negative fifty degrees Fahrenheit, even on a summer's day, our drillers and engineers wear outfits akin to the space suits worn by astronauts on the moon. They live in a comfortable base camp with a wonderful professional chef, and the few times they expose themselves to the elements are when they relax in the pools of hot water created by the AMANDA drills.

In the heroic early days of AMANDA, before a heated, portable hut was built for each drilling site, there were some tough stretches. Teams of ten people sometimes worked up to twenty-four hours without a break—often without gloves when assembling delicate components. But on the whole, the work has been astonishingly unadventurous.

The true challenges have been technical and logistical. Typically we fly 100,000 pounds of cargo and forty people from Christchurch, New Zealand, to the South Pole each summer—enough to fill twenty Hercules C-130 transport planes. It is a massive undertaking, but one accomplished with the utmost efficiency: where Antarctic research is concerned, there is no margin for excess baggage of any kind—be it fiber-optic cable, canisters of fuel or theorists with no real business on the ice.

Once in Antarctica, the operation is orchestrated by Bruce R. Koci, our mechanical engineer and drillmeister. Much of the project is entirely novel, making improvisation the rule, and Koci is a genius at it. Early on, for instance, we found that our hot-water drill was inadequate: it needed eleven days to drill to a depth of 800 meters. So Koci and his colleagues designed a new one. As sleek as a rocket, it dives into the ice gushing 190-degree water from its nose. In its first incarnation, it traveled a thousand meters in four days. These days it goes twice that fast. Mapping AMANDA's geometry is like manufacturing an optical telescope in a dark room. Yet Koci's drill, guided by gravity alone, deviates from the vertical by less than a meter over a depth of two kilometers.

As the drill descends, it leaves a hole about fifty centimeters in diameter, filled with hot water. (Because the hot water is continuously circulated in the hole, and because the ice around it acts like a giant thermos bottle, the water remains liquid for a few days.) Once the drill is removed, the AMANDA crew, often assisted by drillers and volunteers from other scientific missions, attaches a 600-pound weight to a fiber-optic master cable and then drops it into the hole. For the next ten to twenty hours, photomultipliers are attached to the sinking cable with carabiners of the kind used by rock climbers and plugged in at predetermined positions like beads on a rosary. Pressure and temperature meters, lasers, radio receivers, pulsing or steady light-emitting devices and other devices are also attached. (On one occasion a pair of television cameras was sent into the hole; their images can be viewed on our home page at <amanda.berkeley.edu>.)

THEN THE WAITING BEGINS. It takes three to four days for the hole to refreeze completely. Just before the ice turns solid, the pressure spikes dramatically—at a depth of a kilometer, for instance, it rises suddenly from 100 to more than 500 atmospheres. So far, the crib death rate, when the holes refreeze, has hovered around 2 percent. The survivors, encased within their half-inch-thick glass spheres, should live forever—or at least until I die.

As one of the superfluous theorists mentioned above, I have never been invited to Antarctica. But even now, on the nights when drilling goes on there, I can never sleep. To have your career on the line half a world away is hard enough. But to know that you have embroiled so many others in the same improbable adventure, that your funders and colleagues expect results, and that you are totally powerless to affect the outcome, is a form of exquisite torture. And so I keep a laptop at my bedside and check it all through the night for E-mail dispatches.

At 10:30 P.M. on December 24, 1993, when the first of four strings of photomultipliers was deployed and ready for testing, I was at my family's house in Tienen, Belgium, sitting down to a late Christmas Eve dinner. As usual in a Belgian home, the spread was magnificent, but I hardly paid attention. When the news finally arrived, dessert was being served and my laptop was propped on my knees. "First string deployed," the E-mail message read. The sender was too tired to write anything more.

IT WOULD HAVE BEEN EASY to build a more conventional neutrino telescope than AMANDA. We would have covered a square kilometer with spark chambers, shielding them from cosmic radiation with lead plates a few inches thick. The end result would have detected neutrinos beautifully and cost about $10 billion—a thousand times as much as we could afford. Instead, we resigned ourselves to using the simplest instruments possible to detect neutrinos across the greatest possible volume of water at the least possible cost. We would build a telescope that barely works.

Unfortunately, such a design depends, to some extent, on nature's cooperation. So it was that our initial euphoria, on Christmas Eve, turned quickly to perplexity. We knew that some downward-traveling muons, created by cosmic-ray events at the surface, would reach our photomultipliers—even 800 to 1,000 meters beneath the surface. But we detected a hundred times as many as we expected. And though we had expected that bubbles, at that depth, would scatter the Cherenkov light to some degree, what we saw instead was a nearly meaningless blur.

Everything glaciologists had told us about Antarctic ice, it seemed, was wrong. To begin with, the ice down there was far more transparent than anyone had suspected. Condensed from snow that fell 10,000 years ago, at the end of the last ice age, it could transmit a streak of blue light as far as one hundred meters, not just the eight meters that had been predicted. (The discrepancy seemed to arise from the fact that glaciologists carried out their tests with distilled water, which was much less pure than Antarctic ice.) In fact, because our photomultipliers operate at wavelengths where neither atomic nor molecular excitations absorb photons, the ice was almost infinitely transparent. That implied we could detect a few more upward-traveling muons than we had hoped, but it also explained the excess downward-traveling muons.

The bubbles were the real problem, however. A kilometer beneath the surface, we had been told, there would be just a few bubbles, and those would be no more than a micron in diameter. Instead, bubbles were everywhere, and they were fifty times larger than predicted. (We later found a paper on bubbles in ice cores that corroborated our results. But I suspect that some glaciologists still do not believe us.) After a good deal of data analysis and modeling, we predicted that the bubbles would disappear below 1,400 meters. But though we were eventually proved right, we had lost a year by then and still had to go back to drill some new holes.

THIS PAST SUMMER, the second phase of AMANDA was finally calibrated and the third phase was begun. Our telescope is now made up of 420 photomultipliers on thirteen separate strands, sunk between 1,500 and 2,000 meters beneath the Antarctic ice. The photomultipliers are working without a hitch, the telemetry data from the drill and in situ laser measurements agree on their architecture, and the telescope is detecting neutrinos. If the project as a whole has been a roller-coaster ride—one so exciting that we often failed to notice when we were hitting bottom—we are now, undeniably, at a peak. After years of being consumed by engineering problems, we can now at last concentrate on physics. Within days of calibrating the telescope, the intercontinental E-mails that buzz between us changed topics from scattering angle and photomultiplier noise to dark matter, gamma-ray bursts and neutrino oscillations.

Although we have yet to mine three-quarters of our data, we have already begun to follow some veins of interesting science. Theorists have speculated for a decade or so, for instance, that the dark matter in the universe is made up of supersymmetric particles known as neutralinos. According to that theory, most neutralinos are trapped at the centers of stars and planets, where standard instruments cannot detect them. When neutralinos collide at the center of the earth, however, they ought to generate high-energy neutrinos, which AMANDA can detect.

Early on, an investigator in our Stockholm group noted that a muon had skittered up one of AMANDA's strings, illuminating the photomultipliers in sequence like a string of Christmas lights. Of all the neutrinos that had streamed through AMANDA from known cosmic sources, we calculated that only three neutrinos should have aligned randomly with the center of the earth. Yet when we looked at our data, we tentatively identified nine such events. Unfortunately, when we went back over the numbers, the discrepancy between the number of predicted and observed events largely evaporated.

Neutrino oscillation should prove an equally intriguing and elusive quarry. If neutrinos, contrary to current assumptions, are relatively massive—say, between ten and fourteen electron volts—they should oscillate fairly quickly. In that case, even short-range tests, such as the 450-mile neutrino beam that investigators at Fermi National Accelerator Laboratory in Batavia, Illinois, are planning to aim at a neutrino detector in Minnesota, should confirm oscillations. (As it happens, the Fermilab beam passes right under my office in Madison.) If neutrinos have almost no mass, it may take a detector as large as ours, observing neutrinos that come from as far as 10 billion light-years away, to see oscillations. Some of us have even been dreaming about sending a beam from Fermilab to the South Pole: a 6,000-mile trip that may be just long enough to reveal oscillations, if neutrinos have the mass that the Super-K data suggest.

In the next five years we plan to sink seventy more strings of photomultipliers, enabling the telescope to keep watch over a full cubic kilometer of ice. Even then, in some ways, our telescope will be a pretty crude instrument. Compared to a standard optical telescope, its resolution is a joke: as of this writing, it is accurate only to within two degrees of arc—roughly four times the diameter of the full moon—though we hope to cut that margin in half. AMANDA's size alone will give it tremendous power, of course, enabling it to detect gamma-ray bursts and super-massive black holes several times a year. And yet, spectacular though they are, such cosmic light shows should prove no more than opening acts for the main event—whatever that may be.

IN THE PAST, every time astronomers have set their sights on a new wavelength, they have discovered more than they expected. Ricardo Giacconi built an X-ray detector to study solar X rays reflected by the moon, and found neutron and binary stars instead. Karl G. Jansky built an antenna to study shortwave radio interference and discovered radio galaxies. Arno A. Penzias and Robert W. Wilson focused on microwaves and stumbled on the cosmic background radiation, confirming the existence of the big bang. In each case, nature was more imaginative than the people who probed it. If, after nearly twenty years of working on AMANDA, we only discover what we have set out to discover, it will be, in many ways, the most disappointing result imaginable. As Edward W. "Rocky" Kolb, an astrophysicist at Fermilab, put it: "With neutrino astronomy, the real surprise would be that there were no surprises."

Timothy Ferris

Interstellar Spaceflight: Can We Travel to Other Stars?

FROM *SCIENTIFIC AMERICAN PRESENTS*

Despite the starships of science fiction, the vast scale of interstellar space makes traveling across the galaxy just about impossible—at least for human beings. Timothy Ferris, one of our foremost writers on astronomy and the author of Coming of Age in the Milky Way, *proposes a different way to journey to other solar systems.*

Living as we do in technologically triumphant times, we are inclined to view interstellar spaceflight as a technical challenge, like breaking the sound barrier or climbing Mount Everest—something that will no doubt be difficult but feasible, given the right resources and resourcefulness.

This view has much to recommend it. Unmanned interstellar travel has, in a sense, already been achieved, by the Pioneer 10 and 11 and Voyager 1 and 2 probes, which were accelerated by their close encounters with Jupiter to speeds in excess of the sun's escape velocity and are outward-bound forever. By interstellar standards, these spacecraft are slow: Voyager 1, the speediest of the four at 62,000 kilometers per hour (39,000 miles per hour), will wander for several tens of thousands of years before it encounters another star. But the Wright brothers' first airplane wasn't particularly speedy either. A manned interstellar spacecraft that improved on Voyager's velocity by the same 1,000-fold increment by which Voyager improved on the Kitty Hawk flights could reach nearby stars in a matter of decades, if a way could be found to pay its exorbitant fuel bill.

But that's a big "if," and there is another way of looking at the question: Rather than scaling a mountain, one can always scout a pass. In other words, the technical problems involved in traveling to the stars need not be regarded solely as obstacles to be overcome but can instead be viewed as clues, or signposts, that point toward other ways to explore the universe.

Three such clues loom large. First, interstellar space travel appears to be extremely, if not prohibitively, expensive. All the propulsion systems proposed so far for interstellar voyages—fusion rockets, antimatter engines, laser-light sails and so on—would require huge amounts of energy, either in the manufacturing of fusion or antimatter fuel or in the powering of a laser beam for light sails. Second, there is no compelling evidence that alien spacefarers have ever visited Earth. Third, radio waves offer a fast and inexpensive mode of *communication* that could compete effectively with interstellar *travel*. What might these clues imply?

The high cost of interstellar spaceflight suggests that the payloads carried between stars—whether dispatched by humans in the future or by alien spacefarers in the past—are most likely, as a rule, to be small. It is much more affordable to send a grapefruit-size probe than the starship *Enterprise*. Consider spacecraft equipped with laser-light sails, which could be pushed through interstellar space by the beams of powerful lasers based in our solar system. To propel a manned spacecraft to Proxima Centauri, the nearest star, in 40 years, the laser system would need thousands of gigawatts of power, more than the output of all the electricity-generating plants on Earth. But sending a 10-kilogram unmanned payload on the same voyage would require only about 50 gigawatts—still a tremendous amount of power but less than 15 percent of the total U.S. output.

What can be accomplished by a grapefruit-size probe? Quite a lot, actually, especially if such probes have the capacity to replicate themselves, using materials garnered at their landing sites. The concept of self-replicating systems was first studied by mathematician John von Neumann in the 1940s, and now scientists in the field of nanotechnology are investigating how to build them. If the goal is exploring other planetary systems, one could manufacture a few small self-replicating probes and send them to nearby stars at an affordable cost. Once each probe arrived at its destination, it would set up long-term housekeeping on a metallic asteroid. The probe would mine the asteroid and use the ore to construct a base of operations, including a radio transmitter to relay its data back to Earth. The probe could also fashion other probes, which would in turn be sent to other stars. Such a strategy can eventually yield an enormous payoff from a relatively modest investment by providing eyes and ears on an ever increasing number of outposts.

If colonization is the goal, the probes could carry the biological materials required to seed hospitable but lifeless planets. This effort seems feasible whether our aim is simply to promote the spread of life itself or to prepare the way for future human habitation. Of course, there are serious ethical concerns about the legitimacy of homesteading planets that are already endowed with indigenous life. But such worlds may be outnumbered by "near-miss" planets that lack life but could bloom with a bit of tinkering.

One of the intriguing things about small interstellar probes is that they are inconspicuous. A tiny probe built by an alien civilization could be orbiting the sun right now, faithfully phoning home, and we might never learn of its existence. This would be especially true if the probe were engineered to keep a low profile—for instance, if its radio antenna were aimed well away from the ecliptic, or if it were programmed to turn off its transmitters whenever the beam came near a planet. And that is just how such probes would presumably be designed, to discourage emerging species like ours from hunting them down, dismantling them and putting them on display in the Smithsonian National Air and Space Museum. Similarly, a biological probe could have seeded Earth with life in the first place. The fact that life appeared quite early in Earth's history argues against the hypothesis that it was artificially implanted (unless somebody out there was keeping a close eye out for newborn planets), but such an origin for terrestrial life is consistent with the evidence currently in hand.

Where Are the Aliens?

FROM THE SECOND CLUE—that aliens have not yet landed on the White House lawn—we can posit that our immediate celestial neighborhood is probably not home to a multitude of technologically advanced civilizations that spend their time boldly venturing to other star systems on board big, imposing spacecraft. If that were the case, they would have shown up here already, as they evidently have not. (I am, of course, discounting reports of UFO sightings and alien abductions, the evidence for which is unpersuasive.) By similar reasoning we can reach the tentative conclusion that wormholes, stargates and the other faster-than-light transit systems favored by science-fiction writers are not widely in use, at least out here in the galactic suburbs.

Admittedly, one can poke holes in this argument. Perhaps the aliens know we exist but are courteous enough not to bother us. Maybe they visited Earth during the more than three billion years when terrestrial life was all bugs and bacteria and quietly departed after taking a few snapshots and carefully bagging their trash. In any event, it seems reasonable to conclude that if interstellar interstates exist, we are not living near an exit ramp.

The third clue—that radio can convey information much faster and more cheaply than starships can carry cargo—has become well known thanks to SETI, the search for extraterrestrial intelligence. SETI researchers use radio telescopes to listen for signals broadcast by alien civilizations. The SETI literature is therefore concerned mostly with how we can detect such signals and has little to say about how electromagnetic communications might be employed among advanced civilizations as an alternative to interstellar travel. Yet just such a path of speculation can help explain how intelligent life could have emerged in our galaxy without interstellar travel becoming commonplace.

When SETI was first proposed, in a paper published in *Nature* by Giuseppe Cocconi and Philip Morrison in 1959, the main method of electronic communication on Earth was the telephone, and the objection most frequently raised to the idea of interstellar conversation was that it would take too long. A single exchange—"How are you?" "Fine"—would consume 2,000 years if conducted between planets 1,000 light-years apart. But, as Morrison himself has noted, conversation is not essential to communication; one can also learn from a monologue. Eighteenth-century England, for instance, was deeply influenced by the ancient Greeks, although no English subject ever had a conversation with an ancient Greek. We learn from Socrates and Herodotus, although we cannot speak with them. So interstellar communication makes sense even if using it as a telephone does not.

In 1975, when I first proposed that long-term interstellar communications traffic among advanced civilizations would best be handled by an automated network, there was no model of such a system that was familiar to the public. But today the Internet provides a good example of what a monologue-dominated interstellar network might be like and helps us appreciate why extraterrestrials might prefer it to the arduous and expensive business of actually traveling to other stars.

Experientially, the Internet tends to collapse space and time. One looks for things on the Net and makes use of them as one pleases. It does not necessarily matter whether the information came from next door or from the other side of the planet, or whether the items were placed on-line last night or last year. E-mail aside, the Internet is mostly monologue.

Suppose the Internet had been invented several thousand years ago, so that we had access not only to the books of Aristotle and Archimedes but also to their sites on the World Wide Web. What a boon it would be to surf such a web, downloading the lost plays of Sophocles and gazing at the vivid mosaics of Pompeii in colors undimmed by time. Few, I think, would trade that experience for a halting phone conversation with someone from the past.

The same may also be true of communications between alien worlds. The most profound gulf separating intelligent species on various star systems is not space but time, and the best way to bridge that gulf is not with starships but with networked interstellar communications.

The gulf of time is of two kinds. The first is the amount of time it takes signals to travel between contemporaneous civilizations. If, as some of the more optimistic SETI scientists estimate, there are 10,000 communicative worlds in the Milky Way galaxy today, the average time required to send a one-way message to one's nearest neighbor—across the back fence, so to speak—is on the order of 1,000 years. Therefore, it makes sense to send long, fact-filled messages rather than "How are you?"

The Interstellar Internet

THE OTHER GULF arises if, as it seems reasonable to assume, communicative civilizations generally have lifetimes that are brief by comparison with the age of the universe. Obviously, we do not even know whether alien societies exist, much less how long they normally stay on the air before succumbing to decay, disaster or waning interest. But they would have to last a very long time indeed to approach the age of the Milky Way galaxy, which is more than 10 billion years old. Here on Earth, species survive for a couple of million years on average. The Neanderthals lasted about 200,000 years, *Homo erectus* about 1.4 million years. Our species, *H. sapiens,* is about 200,000 years old, so if we are typical, we may expect to endure for another million or so years. The crucial point about any such tenure is that it is cosmologically insignificant. Even if we manage to survive for a robust 10 million years to come, that is still less than a tenth of 1 percent of the age of our galaxy.

Any other intelligent species that learns how to determine the ages of stars and galaxies will come to the same sobering conclusion—that even if communicative civilizations typically stay on the air for fully 10 million years, *only one in 1,000 of all that have inhabited our galaxy is still in existence.* The vast majority belong to the past. Is theirs a silent majority, or have they found a way to leave a record of themselves, their thoughts and their achievements?

That is where an interstellar Internet comes into play. Such a network could be deployed by small robotic probes like the ones described earlier, each of which would set up antennae that connect it to the civilizations of nearby stars and to other network nodes. The network would handle the interstellar radio traffic of all the worlds that know about it. That would be the immediate payoff: one could get in touch with many civilizations, without the need to establish contact with each individually. More important, each node would keep and distribute a record of the data it handled. Those records would vastly enrich the network's value to every civilization that uses it. With so many data constantly circulated and archived among its nodes, the interstellar Internet would give each inhabited planet relatively easy access to a wealth of information about the civilizations that currently exist and the many more worlds that were in touch with the network in the past.

Intelligence brings knowledge of one's own mortality—and at the same time, provides a means to transcend it—so the desire for some kind of immortality is, I suspect, widespread among intelligent beings. Although some species may have limited themselves to physical monuments, such as the one erected by Percy Bysshe Shelley's Ozymandias, these must eventually weather away and would in any event require long journeys to be seen and appreciated. Surely most species would elect to contribute to the interstellar Internet, where their thoughts and stories could career around the galaxy forever.

If there were any truth in this fancy, what would our galaxy look like? Well, we would find that interstellar voyages by starships of the *Enterprise* class would be rare, because most intelligent beings would prefer to explore the galaxy and to plumb its long history through the more efficient method of cruising the Net. When interstellar travel did occur, it would usually take the form of small, inconspicuous probes, designed to expand the network, quietly conduct research and seed infertile planets. Radio traffic on the Net would be difficult for technologically emerging worlds to intercept, because nearly all of it would be locked into high-bandwidth, pencil-thin beams linking established planets with automated nodes. Our hopes for SETI would rest principally on the extent to which the Net bothers to maintain omnidirectional broadcast antennae, which are economically draining but could from time to time bring in a fresh, naive species—perhaps even one way out here beyond the Milky Way's Sagittarius Arm. The galaxy would look quiet and serene, although in fact it would be alive with thought.

In short, it would look just as it does.

STEPHEN S. HALL

Journey to the Center of My Mind

FROM THE *NEW YORK TIMES MAGAZINE*

With new approaches and new technology, the once uncharted landscape of the brain is slowly being mapped. Magnetic resonance imaging lets scientists observe which areas of the brain are used while a human subject performs different tasks. Cheerfully taking on the role of a test subject, Stephen S. Hall sends an enthusiastic dispatch from this frontier.

A t a few minutes after 4:00 on a Sunday afternoon in January, when most of New York was tuning in to the playoff game between the Jets and the Broncos, I had something else on my mind. More precisely, I had something around my mind, namely the 1.5-ton magnet of the magnetic resonance imaging (M.R.I.) machine at the Memorial Sloan-Kettering Cancer Center. I wasn't there for medical reasons; I was there to embark on an adventure. My journey would take me no farther than this laboratory on the Upper East Side of Manhattan, and yet I was going somewhere very few people have been.

Of all the frontiers that await exploration, perhaps no other is more intriguing than the terra incognita that lies between our ears. There, in a three-pound pudding of neurons and wiring, lie the keys to the kingdoms of memory, of thought, of desire, of fear, of the habits and skills that add up to who we each are. It is an especially daunting frontier because even after you have entered the realm of the brain, it's still necessary to locate a second, far more elusive boundary that separates the mere hardware of neurology from that elusive quality known as the mind, the "I" that hovers in the background of all conscious mental activity. I had hopes of getting a glimpse of that I—my mind—in the course of my travels.

My Virgil on this journey through the dark wood of cognition was Joy Hirsch, a voluble, cheerful scientist with dark bangs over her forehead and inch-long nails. Hirsch directs the Functional Imaging Laboratory at Sloan-Kettering and is a professor of neuroscience at the Weill Medical College of Cornell University. I'd become interested in her work several years ago when her group published a paper in *Nature* suggesting that you could tell whether a person learned a foreign language early or late in life by pinpointing the exact location of the speech center in the brain. When I proposed taking a tour of my own brain, Joy saw it as an opportunity to extend her research into some new areas.

The premise was simple. I would undergo a series of brain-imaging sessions using the technology of functional M.R.I.—like diagnostic M.R.I., except that it also measures brain activity. These scans would, for the most part, be customized, almost autobiographical studies that would probe thoughts and emotions related to my personal history and work as a writer and editor. My journey, I understood, would be unscientific—no study of one individual holds statistical significance. But even as a neural picaresque, it wasn't without value. We designed exercises with reasonable experimental controls and, whenever possible, in ways that might complement published studies. We set out to investigate, among other things, the use of figurative language, the neural residue of emotional memories, the seat of humor, the source of sentence composition and even the cognitive headwaters of storytelling.

Before setting out, I consulted several prominent brain scientists; Steven Pinker, the cognitive scientist at M.I.T., provided the most sage advice about the proper frame of mind for doing experiments in an M.R.I. machine. "Focus is essential," he said. "You almost have to be a Zen master." Easier said than done.

The Lay of the Land

FOR THE FIRST SESSION, Joy Hirsch set up fourteen scans, designed to sketch out a rough map of the auditory, visual, touch, motor and language systems of my brain. "This is like a tour of the building before we start to talk to the individual departments," Joy said. This kind of mapping was pioneered by Hirsch's team at Sloan-Kettering to provide a guide for neurosurgeons so that they can avoid speech, hearing and other critical centers when removing brain tumors. (One condition for my participation was that patient scans always took precedence.)

As I lay flat on my back, about to enter the narrow bore of the scanner for the first time, Joy and a technician, Greg Nyman, sandwiched my head between cushions, placed a piece of tape across my forehead and slipped a plastic cowl over my head. "We're going to put you in," Joy announced as I felt myself slide about four feet into an aperture 23 inches wide and only 17 inches high. Staring up at a mirror above my head, I could look out of the tube, where the vista included two familiar shoes forming a nervous V in the air. Each "run," Joy explained, would last 144 seconds; the machine divided my brain into about 185,000 units, or voxels, and measured the activity in each every four seconds. The huge, room-size magnet allows the machine to detect subtle changes in blood flow in each voxel, changes that are believed to reflect levels of brain activity.

Science is never quite as seamless as it appears in the pages of journals. The first few runs went fine—a flashing checkerboard pattern excited my visual cortex, as planned, and Joy rubbed an ordinary five-and-dime pot-scrubber over my right hand, to stimulate the tactile (or somatosensory) part of my brain. I had great difficulty seeing and hearing, however. I couldn't wear my glasses inside the machine—the powerful magnetic field would turn them into a ballistic missile. The prescription goggles on hand in the lab provided just enough visual acuity to turn an exercise where I named objects into a Magoo-ish misadventure. I mistook a tennis racket for a globe and a canoe for a comb. Later, straining to hear words through earphones, I could make out only a couple over the din of the machine, which sounds like the loudest, most emphatic busy signal imaginable. Again, I resorted to pure conjecture. At this rate, I began to fear we were using our precious M.R.I. time to pinpoint nothing more than the neural headquarters for Guesswork.

Finally, we made our first tentative sortie into the land of cognition. Joy knew I spoke Italian, and I was curious to see where the "Italian speech module" was located in my brain. I had learned the language in my midtwenties while living in Rome, but I thought there might be a chance I'd picked up a smattering of Italian as a child, since my maternal grandparents spoke it almost exclusively. It turns out that all languages learned early in life cohabit the same neural real estate, whereas a foreign language learned as an adult usually occupies a distinctly separate region. Would the conversations in Italian I had overheard as a child have left a neural residue?

To find out, Joy asked me to perform a task that formed the basis of the group's 1997 *Nature* paper. I had 10 seconds to tell a little story in my head in response to a series of visual prompts—a picture showing either sunrise, noon or night. During the first run, I would think up a scenario in English; the second time, in Italian. The morning scene, for example, elicited the following: "I woke up around eight o'clock, had a bite to eat, put on my coat and walked to the subway, which I took to work." All I can say in my literary defense is, you try being clever with a deadline of 10 seconds and 110-decibel honks in your ears about five times a second.

Lying stock still in a horizontal phone booth might not sound like much of a physical adventure, but after fifty minutes I felt exhausted by the effort of inactivity. You're supposed to keep your head as steady as a statue, and just as empty. But any active, imaginative intelligence is apt to daydream, worry, have idle thoughts—all forms of mental static. Thoughts kept crashing through the artificial quietude of my empty head. "What's the Italian word for joke?" "Wonder how the Jets are doing?" "Man, is my mouth dry!" Theoretically, each of those errant thoughts has its latitude and longitude, its little magnetic wrinkle destined, perhaps, to show up in the raw data and fog this high-tech mirror in which I hoped to see something of myself.

All Roads Lead to . . . Broca's Area

WHEN I MET with Joy to go over the results from the first session, we sat at a round table in her office, with twenty-one cross-sectional slices of my brain up on the light box, a brain atlas open on the table and small printouts of each cross-section in front of her, marked by yellow Post-it arrows pointing to the predominant landmarks of my neural anatomy. She had carefully compared the lay of my brain to a standard atlas and, slice by slice, walked me through the basic landscape. "I love looking at brains," she confided at one point. "They're very beautiful, very intimate." Everything looked normal. "This is a beautiful, textbook brain," she said with obvious enthusiasm. But I almost didn't hear her. I was mesmerized by the beauty of this hidden landscape.

Now that I was on a guided tour of my own brain, I began to appreciate just how precise the convoluted geography of the brain actually is. Like a lot of laypeople, I thought the whorls and folds on the surface layer of the brain, known as the cortex, varied by individual like so many fingerprints. In fact, the patterns are basically the same in everyone. As I could see in Joy's atlas, each bulge (or gyrus) and each crevice (or sulcus) is as precisely plotted as any topographic map. All the cognitive action happens in those whorls; everything else is scaffolding, underground cable, antique structures handed down by evolution from reptile and early mammalian brains.

The first departments we checked were the somatosensory and motor regions. The touch of the pot-scrubber caused a small section of the post-central gyrus on the left side to light up, as expected, while the finger-tapping exercise caused a herd of neurons just across the deep neural ravine from the tactile center to become active.

"This is very typical," Joy said. "In fact, it's particularly nice. Some brains have a little bit less specific activity. One of the nice things about the activity patterns in your brain is that they are very localized. Your brain just goes and gets the job done. It doesn't waste a lot of energy going other places."

I'd been concerned about my inability to see and hear stimuli during the first session, but those difficulties actually made the results more interesting. When I was straining to make out images during the object-naming task, my visual cortex looked like the wall of napalm scene in *Apocalypse Now*. "I'll tell you, this little brain was working mighty hard to get that information," Joy said with a laugh. The same evidence of effort showed up in the auditory map—the place in the midbrain that "listens," called the transverse temporal gyrus, was screaming in technicolor, even though I could only make out one or two words.

Finally, we looked at my basic language centers. Language, of course, is normally located on the dominant side of the brain, which in right-handers like me is on the left side. When I spoke to myself in English, therefore, a small patch of cortex on the left side lighted up. This is known as Broca's area, after the French pathologist who first identified it.

The data from the Italian-speaking exercise unfortunately were ambiguous. The image was filled with green blotches that signified movement. As anyone who watched Roberto Benigni at the Academy Awards knows, Italian is a particularly kinetic language. "The only time you moved your head the whole time was when you were speaking Italian!" Joy cried, explaining why she couldn't definitively say if my Italian was learned early or late. With prodding, she pointed out a tiny gap between the areas that lighted up when I spoke to myself in English and Italian. "If I had to call it," she said, "you have Italian sitting above the English, and both of them are pinned together here in Broca's area." That tiny gap, about five millimeters, suggested that I'd learned Italian as an adult, not as a child.

If there was a lesson from this first expedition, it was this: in terms of neural architecture, we all live in Levittown. Every brain is pretty much like the next. What makes each unique is how we decorate them, as it were, with experience and memory and habits and skills. Staring at these voluptuous, serpentine folds of cortex, I was struck by the strange commingling of inert anatomy and transcendent human qualities. Buried in those headlands and crevices, I knew, were mental images of grandparents no longer alive, of my mother trying to explain death to me for the first time, the sound of loved ones' voices, my father encouraging me as we played catch, of Roberto Clemente whirling and throwing a baseball, as well as the state capital of Vermont, the square root of 81 and the narrative line of three books I've written. The vastness and steadfastness of those memories, all nestled and synaptically etched in this bland gray and squishy landscape, was a miracle impossible to capture on film and perhaps beyond the grasp of our very modest experiments.

But we could try. We decided to devote the next session to exploring things that might be unique to me as a writer and as a person. We would search for the headwaters of storytelling, and we'd try to see if the brain reacted differently when it encountered family and friends as opposed to the faces and voices of strangers.

To the Source of the Narrative

JOY AND DIANA MORENO, an Argentine graduate student in her lab, devised a plan to watch my brain in action as it invented a story. They would provide me with a set of narrative prompts—either a sequence of simple images (like a dog or a tree) or, using a different sensory route to the same destination, a sequence of distinctive sound effects (a honking horn, a crying baby). While the M.R.I. was blasting away, I'd ad-lib a running story in my head.

The results were fascinating. We saw the visual cortex light up as expected. But we also saw many small, discrete precincts of the frontal lobe activated on the right side of the brain, with particularly intense activity in an area called the inferior frontal gyrus—what might be called the storytelling area. "This is not a subtle effect, Steve," Joy pointed out. "In this business, this is big." It seemed to cover at least 1.5 cubic centimeters—about the size of a sugar cube. I asked her what might account for its size, and she replied that it could be a combination of things: a natural predisposition to use the right side of my brain (nature) and probably my frequent use of it (nurture) in the course of a lifetime.

There were other areas involved; they formed a network, actually. But Joy seemed particularly taken with the notion that the brain parceled out two related capabilities—one to create stories and the other to articulate them in speech—in essentially the same place in the two hemispheres, for my storytelling area was the mirror image of the site on the left, dominant side of the brain that controls speech.

Next we compared familiar voices and faces to unfamiliar voices and faces. When I had suggested the idea of looking at familiar faces while in the machine, I thought it might serve as a back road to memory and feeling; the difference between the perception of an unknown and known face should be the neural residue of recognition and familiarity. Once again, we were in for some surprises.

As the M.R.I. beeped around my head, I was shown a series of fifty-two photographs over the course of four runs, thirteen per run. During the first two runs, none of the faces were familiar—or, at least, they weren't supposed to be. One picture happened to be a spitting image of the girl—I'll call her D.—I had a crush on in high school and whom I'd asked, unsuccessfully, to the senior prom. As I processed this image, I experienced what felt like an Etna of neural activity somewhere in my cranium.

During the next two runs, I was shown only familiar faces—indeed, painfully familiar in the case of images of myself as Sullen Adolescent and Alienated Hirsute Expatriate. My wife, Mindy, had sent Joy's lab photographs of practically everybody I knew: my parents, Mindy giving birth to our son, my nephews, the college friend I bummed around with in Europe, old family friends, even my old landlady from Rome, as well as six photos of my younger self, including a high-school graduation picture of me looking so clueless that I wouldn't have gone to the prom with me.

When we reviewed the brain scans a few days later, I had the thrill of witnessing a unique feature of the physical world for the first time. During the familiar-faces experiment, we saw activity in the visual cortex; no surprise there. We saw activity in the hippocampus. No surprise there, either; the hippocampus, a structure deep in the interior of the brain, is thought to be involved in the storage of long-term memory. "Your whole hippocampus is screaming!" Joy said. We also saw activity in a structure adjacent to the hippocampus known as the fusiform gyrus; this too, was not a surprise, at least not to Joy. Recent research on face recognition has identified this as the key area in the brain for the specialized task of perceiving faces. What was a surprise was that the most excited sector in my brain as it viewed familiar faces was, once again, the "storytelling area."

Next, we performed the same exercise using familiar voices: relatives and friends provided the lab with taped monologues addressed to me. Unfortunately, I had the same difficulty making out the voices as in the earlier session; the only voice I could identify unambiguously was that of my wife. I guessed—correctly, as it turns out—that other voices belonged to my father, mother and daughter, although I didn't learn until later what they had actually said. My father recalled going to the 1966 All-Star game in St. Louis and meeting the Cub infielder Ron Santo on the plane home. My mother invoked memories of what she called the "friendly persuader" of my mischievous childhood (the wooden spoon). And my three-year-old daughter, Micaela, had chattered, "Testing, one-two-three, Daddy, I love you, testing one-two-three." I didn't hear any of it.

But, remarkably, my brain apparently did. The scans showed the exact same far-flung network of parts fired up as when I was looking at familiar pictures—including the storytelling region in my right brain. It was almost as if I subconsciously filtered out the background noise and heard what I needed to hear. And not only that, simply hearing familiar voices activated the visual cortex, as if I mentally pictured the people whose voices I heard.

Joy used that all-purpose, noncommittal, scientifically discreet word to describe this effect: interesting. Exercising the layperson's right to speculate, I was immediately taken by a literary notion: the role of narrative in memory. What the data seemed to suggest to me was that part of what makes a face or a voice familiar to us is that the brain attaches it to a narrative. Perhaps we "tag" people with narratives to help us remember them. The image or voice (and perhaps even the taste, touch or smell) of a familiar person summons up from our memories the story we've woven them into. That madeleine of Marcel Proust's seems less like a literary conceit and more like a brilliant scientific insight—but then Proust understood at least as much about memory as any modern neuroscientist.

The Land of Metaphor

TO MAP THE CREATION of sentences in my brain, Joy showed me a series of words—highway, dawn, border—and asked me to use them in the most elaborate sentences I could think of, all in ten seconds. One that stuck in my mind was, "The border between here and there is uncertain, and moving further into the future even as we speak."

Joy was particularly amused when she showed me the results of this exercise. The areas of activity were small, relatively discrete and, unlike virtually all of the other creative tasks I performed, well represented on the left as well as the right side of the brain, indicating, she said, excellent mental economy.

I reverted to my right-hemisphere bias on two other tasks: creating metaphors and synonyms. I suppose there might be something socially redeeming about spitting out a metaphor on cue, every four seconds, but after shoveling out ten examples of figurative language in forty seconds, it began to feel like an aerobic activity. The cues, once again, were simple illustrations: a globe, a wreath, a canoe, a snail, a tennis racket, a seahorse and so on. Considering some of my responses—"Up the creek without a paddle" and "You've got quite a racket going on," to name two—I worried that we'd overshot Metaphor Mountain on the map and landed in the Slough of Cliche.

Not so. During this task—and to a great degree, during the Great Thesauric Expedition (as I now call the synonym task)—the right side of my brain lighted up like a neon sign on a cheap diner. Almost all the usual suspects were on display: parts of the visual cortex, the language area, that interesting storytelling area in the inferior frontal gyrus and a spot toward the top of the brain, the medial frontal gyrus, that Joy believed was organizing and coordinating high-level activity on a number of tasks. With Joy's help, I was beginning to recognize a network I seemed to use over and over again.

As we reviewed the results, it occurred to me that we had begun to exhaust the usefulness of the geographic metaphor—and perhaps that was the real point of the entire exercise. The more complex the task, the more dispersed the brain's activity. The pattern in the scans stopped looking like a landscape with a few isolated peaks and more like a circuit with an extravagant number of relay points.

The potential link between circuitry and consciousness became especially clear when we went looking for the seat of humor. Joy showed me Gary Larson cartoons, first with neutral, unfunny captions and then with their proper punch lines. Perhaps because of the circumstances, one cartoon in particular had me struggling to suppress a laugh. It showed a group of doctors in the midst of brain surgery. The neutral caption read, "Operating Room"; the Larson caption had one of the surgeons exclaiming: "Wow! His brain still uses vacuum tubes!"

In response to this and other cartoons, my brain looked like those aerial shots of Southern California during brush-fire season: there were little embers of neural activity all over. The hippocampus lighted up, suggesting the involvement of memory; the thalamus on the right side became active (the first time we'd seen that in any of our experiments), suggesting sensory processing, and we even detected a little activity in the sensorimotor cortex, which normally controls physical movement. Joy immediately thought "smile," and I thought "laugh." Like a Chopin impromptu, my "humor network" hit a great many notes, high in the brain and low, and did so with lightning rapidity: visual processing, language processing, memory, the perception of a cognitive disjunction, and all of it seemingly wired to trip a laugh instantaneously. Moreover, this network began to suggest something more complex than mere cognition—something like consciousness, for humor is very personal, turning as it does on such idiosyncratic traits as one's sense of irony, cognitive dissonance and *Schadenfreude*. The network we were seeing, with its unique linkages, represented my sense of humor.

There Is No Center There

AS I LOOK at my brain again, slice by slice, holding the film up to a window in my study, I am struck once more with the everyday wonder of the landscape—the shadowy lines of sulci running like streams to nowhere from the interior of the cerebrum, the peninsular gyri, each plump with purpose and secrets, and, like a river running through it all, the midline separating right hemisphere from left. And more than ever, I realize that the organizing metaphor for this expedition—a journey to the center of my mind—has been misleading. As task after task demonstrated, there is no center of activity, only way stations in a circuit, winking at each other in milliseconds, churning in some mysterious neural communion. And the notion of mind? We didn't make much progress penetrating that mystery. Perhaps it's nothing more than the heat given off by our personalized circuits, everywhere and yet nowhere.

If, as Joseph Conrad once said, the most interesting places are the empty spaces on a map, the prefrontal cortex must be an especially fascinating place for brain scientists. We were puzzled by the general dearth of activity in my frontal lobe; although it is supposed to be the real crucible of human thought, none of our exercises seemed to tickle it into much activity. Indeed, there were many questions I wished to ask that we couldn't approach for reasons of practicality or time, or an inability to even formulate a workable experimental question. I was interested, for example, in exploring skills more particular to an editor than a writer, like fixing ungrammatical sentences. I would have liked to probe emotions more intently, especially things we feel every day like anger and insecurity and sexual arousal, but it turns out they are exceedingly difficult to test in a meaningful way. I would have also liked to see what the brain looked like as it wrestled with a moral dilemma.

Brain scientists would like to know the answers to many of those questions, too. But as Joy, whose great-grandmother traveled the Oregon Trail, likes to point out, we're still in the "covered wagon" phase of mapping brain function. "When the real pioneers started their journey, they had no shortage of ideas about what Oregon was going to look like." she said. "And similarly, we have no shortage of ideas about mind and consciousness, even though we really haven't gone very far into the frontier." Even a modest journey like mine, however, hints at the territory ahead. "We have been able to observe several of the interconnected systems in your brain and perhaps have glimpsed some of your consciousness at work, even if it was only a snapshot of a brief instant of your life."

In our age-old struggle to understand the mind, we have always been empowered—yet oddly constrained—by the vocabulary of the moment, be it the voices of the gods in ancient myth, buried conflicts in the idiom of Freudian analysis or associative memories in Proustian terms. But as psychology and neuroscience begin to converge, brain imaging may actually provide a new, visual vocabulary with which to rethink, and perhaps reconcile, some of these older ideas of mind. A common thread of both the Freudian and Proustian worldviews is the associative quality of recollection—the odd word or sight that connects to a deeper trauma, the odor that connects to a more extensive memory. Association requires connections, and as I saw, a brain scan of humor, for example, can actually depict a rich skein of associations in a diagram of neural connections. Preposterous as it may seem, I can imagine a day in the distant future when the M.R.I. replaces the couch, when the therapist uses words or odors or pictures to excite and pinpoint circuitry and then the neuroanatomist translates the images into explanations of behavior. Of course, there is always the possibility that after decades of exploration in search of mind, we'll still find ourselves, metaphorically speaking, knee-deep in a swamp of neurotransmitters that may bring us no closer to a biological understanding of "mind."

It's odd to put it this way, but I may know more about how my brain works than almost any human who has ever lived, and yet that knowledge has won nothing more than a beachhead on a vast, uncharted continent. That is no small achievement, though my journey makes clear that these are early days in the brain-mapping business. I cannot say what in my genetics or upbringing might have contributed to the hive of activity we observed in the right side of my brain. Or why I remember everything about the moment when my friend D. agreed to go to the prom with me thirty years ago but nothing about the moment when she changed her mind. Given the insecurities of adolescence and the uncertainties of affection, it seemed at the time like one of those watershed moments of my life, and even now it can still produce a wince, but it appears to have eluded the gaze of the M.R.I. machine. For when we went back and examined the scans taken in the moment when I thought I'd been shown D.'s face, the data had been subsumed in a stream of average responses. It didn't leave a trace.

Floyd Skloot

Gray Area: Thinking with a Damaged Brain

FROM *CREATIVE NONFICTION*

Despite the new vistas that brain research is opening, there is still much about the brain that is not understood, and so far scientists and doctors have had little to offer people who suffer from some form of brain damage. A poet and a writer, Floyd Skloot contracted a virus that attacked his brain and severely affected his memory, reasoning abilities and cognitive powers—elements essential not only to everyday living but to a true sense of self. In this unflinching and poignant essay, he describes his efforts to come to terms with the new person he has become.

I used to be able to think. My brain's circuits were all connected and I had spark, a quickness of mind that let me function well in the world. There were no problems with numbers or abstract reasoning; I could find the right word, could hold a thought in mind, match faces with names, converse coherently in crowded hallways, learn new tasks. I had a memory and an intuition that I could trust.

All that changed on December 7, 1988, when I contracted a virus that targeted my brain. A decade later, my cane and odd gait are the most visible evidence of damage. But most of the damage is hidden. My cerebral cortex, the gray matter that M.I.T. neuroscientist Steven Pinker likens to "a large sheet of two-dimensional tissue that has been wadded up to fit inside the spherical skull," has been riddled with tiny perforations. This sheet and the thinking it governs are now porous. Invisible to the naked eye, but readily seen through brain imaging technology, are areas of scar tissue that constrict blood flow. Anatomic holes, the lesions in my gray matter, appear as a scatter of white spots like bubbles or a ghostly pattern of potshots. Their effect is dramatic; I am like the brain-damaged patient described by neuroscientist V. S. Ramachandran in his book *Phantoms in the Brain:* "Parts of her had forever vanished, lost in patches of permanently atrophied brain tissue." More hidden still are lesions in my Self, fissures in the thought process that result from this damage to my brain. When the brain changes, the mind changes—these lesions have altered who I am.

"When a disease process hits the brain," writes Dartmouth psychiatry professor Michael Gazzaniga in *Mind Matters,* "the loss of nerve cells is easy to detect." Neurologists have a host of clinical tests that let them observe what a brain-damaged patient can and cannot do. They stroke his sole to test for a spinal reflex known as Babinski's sign or have him stand with feet together and eyes closed to see if the ability to maintain posture is compromised. They ask him to repeat a set of seven random digits forward and four in reverse order, to spell *world* backwards, to remember three specific words such as *barn* and *handsome* and *job* after a spell of unrelated conversation. A new laboratory technique, positron emission tomography, uses radioactively labeled oxygen or glucose that essentially lights up specific and different areas of the brain being used when a person speaks words or sees words or hears words, revealing the organic location for areas of behavioral malfunction. Another new technique, functional magnetic resonance imaging, allows increases in brain blood flow generated by certain actions to be measured. The resulting computer-generated pictures, eerily colorful relief maps of the brain's lunar topography, pinpoint hidden damage zones.

But I do not need a sophisticated and expensive high-tech test to know what my damaged brain looks like. People living with such injuries know intimately that things are awry. They see it in activities of daily living, in the way simple tasks become unmanageable. This morning, preparing oatmeal for my wife, Beverly, I carefully measured out one-third cup of oats and poured them onto the pan's lid rather than into the bowl. In its absence, a reliably functioning brain is something I can almost feel viscerally. The zip of connection, the shock of axon-to-axon information flow across a synapse, is not simply a textbook affair for me. Sometimes I see my brain as a scalded pudding, with fluky dark spots here and there through its dense layers and small scoops missing.

Sometimes I see it as an eviscerated old TV console, wires all disconnected and misconnected, tubes blown, dust in the crevices.

Some of this personal, low-tech evidence is apparent in basic functions like walking, which for me requires intense concentration, as does maintaining balance and even breathing if I am tired. It is apparent in activities requiring the processing of certain fundamental information. For example, no matter how many times I have been shown how to do it, I cannot assemble our vacuum cleaner or our poultry shears or the attachments for our hand-cranked pasta maker. At my writing desk, I finish a note and place the pen in my half-full mug of tea rather than in its holder, which quite obviously teems with other pens. I struggle to figure out how a pillow goes into a pillowcase. I cannot properly adjust Beverly's stereo receiver in order to listen to the radio; it has been and remains useful to me only in its present setting as a CD player. These are all public, easily discernible malfunctions.

However, it is in the utterly private sphere that I most acutely experience how changed I am. Ramachandran compares this to harboring a zombie, playing host to a completely nonconscious being somewhere inside yourself. For me, being brain-damaged also has a physical, conscious component. Alone with my ideas and dreams and feelings, turned inward by the isolation and timelessness of chronic illness, I face a kind of ongoing mental vertigo in which thoughts teeter and topple into those fissures of cognition I mentioned earlier. I lose my way. I spend a lot of time staring into space, probably with my jaw drooping, as my concentration fragments and my focus dissolves. Thought itself has become a gray area, a matter of blurred edges and lost distinctions, with little that is sharp about it. This is not the way I used to be.

In their fascinating study, "Brain Repair," an international trio of neuroscientists—Donald G. Stein from America, Simon Brailowsky from Mexico and Bruno Will from France—report that after injury "both cortical and subcortical structures undergo dramatic changes in the pattern of blood flow and neural activity, even those structures that do not appear to be directly or primarily connected with the zone of injury." From this observation, they conclude that "the entire brain—not just the region around the area of damage—reorganizes in response to brain injury." The implications of this are staggering; my entire brain, the organ by which my very consciousness is controlled, was reorganized one day ten years ago. I went to sleep *here* and woke up *there;* the place looked the same, but nothing in it worked the way it used to.

If Descartes was correct, and to Think is to Be, then what happens when I cannot think, or at least cannot think as I did, cannot think well enough to function in a job or in the world? Who am I?

You should hear me talk. I often come to a complete stop in midsentence, unable to find a word I need, and this silence is an apt reflection of the impulse blockage occurring in my brain. Sitting next to Beverly as she drives our pickup truck through Portland traffic at 6 P.M., I say, "We should have gone for pizza to avoid this blood . . ." and cannot go on. I hear myself; I know I was about to say "blood tower traffic" instead of "rush hour traffic." Or I manifest staggered speech patterns—which feels like speaking with a limp—as I attempt to locate an elusive word. "I went to the . . . *hospital* yesterday for some . . . *tests* because my head . . . *hurt.*" Or I blunder on, consumed by a feeling that something is going wrong, as when I put fresh grounds into the empty carafe instead of the filter basket on my coffee maker, put eye drops in my nose or spray the cleaning mist into my face instead of onto the shower walls. So at the dinner table I might say "Pass the sawdust" instead of "Pass the rice," knowing even as it happens that I am saying something inappropriate. I might start a conversation about "Winston Salem's new CD" instead of Wynton Marsalis' or announce that "the shore is breaking" when I mean to say "the shower is leaking." There is nothing smooth or unified anymore about the process by which I communicate; it is disintegrated and unpredictably awkward. My brain has suddenly become like an old man's. Neurologist David Goldblatt has developed a table which correlates cognitive decline in age-associated memory impairment and traumatic brain injury, and the parallels are remarkable. Not gradually, the way such changes occur naturally, but overnight, I was geezered.

It is not just about words. I am also *dyscalculic*, struggling with the math required to halve a recipe or to figure out how many more pages are left in a book I'm reading. If we are on East 82nd and Third Avenue in Manhattan, staying with my childhood friend Larry Salander for the week, it is very difficult for me to compute how far away the Gotham Book Mart is over on West 47th between Fifth and Sixth, though I spent much of my childhood in the city.

Because it is a place where I still try to operate normally, the kitchen is an ideal neurological observatory. After putting the leftover chicken in a plastic bag, I stick it back in the oven instead of the refrigerator. I put the freshly cleaned pan in the refrigerator, which is how I figure out that I must have put the chicken someplace else because it's missing. I pick up a chef's knife by its blade. I cut off an eighth of a giant white onion and then try to stuff the remainder into a recycled 16-ounce yogurt container that might just hold the small portion I set aside. I assemble ingredients for a vinaigrette dressing, pouring the oil into an old olive jar, adding balsamic vinegar, mustard, a touch of fresh lemon juice and spices. Then I screw the lid on upside-down and shake vigorously, spewing the contents everywhere. I stack the newspaper in the wood stove for recycling. I walk the garbage up our 200-yard-long driveway and try to put it in the mailbox instead of the trash container.

At home is one thing; when I perform these gaffes in public, the effect is often humiliating. I can be a spectacle. In a music store last fall, I was seeking an instruction book for Beverly, who wanted to relearn how to play her old recorder. She informed me that there were several kinds of recorders; it was important to buy exactly the right category of book since instructions for a soprano recorder would do her no good while learning on an alto. I made my way up to the counter and nodded when the saleswoman asked what I wanted. Nothing came out of my mouth, but I did manage to gesture over my right shoulder like an umpire signaling an out. I knew I was in trouble, but forged ahead anyway, saying, "Where are the books for sombrero reporters?" Last summer in Manhattan, I routinely exited the subway stations and led Beverly in the wrong direction, no matter which way we intended to go. She kept saying things like, "I think west is *that* way, sweetie," while I confidently and mistakenly headed east, into the glare of the morning sun, or, "Isn't that the river?" as I led her away from our riverside destination. Last week, in downtown Portland on a warm November morning, I stopped at the corner of 10th and Burnside, one of the busiest crossings in the city, carefully checked the traffic light (red) and the traffic lanes (bus coming), and started to walk into the street. A muttering transient standing beside me on his way to Powell's Books, where he was going to trade in his overnight haul of tomes for cash, grabbed my shoulder just in time.

At home or not at home, it ultimately makes no difference. The sensation of *dysfunctional mentation* is like being caught in a spiral of lostness. Outside the house, I operate with sporadic success, often not knowing where I am or where I'm going or what I'm doing. Inside the house, the same feelings often apply and I find myself standing at the top of the staircase wondering why I am going down. Even inside my head there is a feeling of being lost, thoughts that go nowhere, emptiness where I expect to find words or ideas, dreams I never remember.

Back in the fall, when it was Beverly's birthday, at least I did remember to go to the music store. More often, I forget what I am after within seconds of beginning the search. As she gets dressed for work, Beverly will tell me what she wants packed for lunch and I will forget her menu by the time I get up the fourteen stairs. Now I write her order down like a waiter. Sometimes I think I should carry a pen at all times. In the midst of preparing a salad, I stop to walk the four paces over to the little desk where we keep our shopping list and forget "tomatoes" by the time I get there. So I should also have paper handy everywhere. Between looking up a phone number and dialing it, I forget the sequence. I need the whole phone book on my speed-dial system.

Though they appear without warning, these snafus are no longer strange to me. I know where they come from. As Dr. Richard M. Restak notes in *The Modular Brain,* "A common error frequently resulting from brain damage involves producing a semantically related word instead of the correct response." But these paraphasias and neologisms, my *expressive aphasias,* and my dyscalculas and my failures to process—the rapids of confusion through which I feel myself flailing—though common for me and others with brain damage, are more than symptoms to me. They are also more than what neurologists like to call *deficits,* the word of choice when describing impairment or incapacity of neurological function, as Oliver Sacks explains in his introduction to *The Man Who Mistook His Wife for a Hat.* These "deficits" have been incorporated into my very being, my consciousness. They are now part of my repertoire. Deficits imply losses; I have to know how to see them as gains.

PRACTITIONERS OF NEUROSCIENCE call the damage caused by trauma, stroke or disease "an insult to the brain." So pervasive is this language that the states of Georgia, Kentucky and Minnesota, among others, incorporate the phrase "insult to the brain" in their statutory definitions of traumatic brain injury for disability determinations. Such insults, according to the Brain Injury Association of Utah, "may produce a diminished or altered state of consciousness, which results in an impairment of cognitive abilities or physical functioning." The death of one Miles Dethmuffen, front man and founding member of the Boston rock band Dethmuffen, was attributed in news reports to "an alcoholic insult to the brain." The language used is so cool. There is this sentence from the Web site NeuroAdvance.com: "When there is an insult to the brain, some of the cells die." Yes.

Insult is an exquisitely zany word for the catastrophic neurological event it is meant to describe. In current usage, of course, insult generally refers to an offensive remark or action, an affront, a violation of mannerly conduct. To insult is to treat with gross insensitivity, insolence or contemptuous rudeness. The medical meaning, however, as with so many other medical words and phrases, is different, older, linked to a sense of the word that is some two or three centuries out of date. *Insult* comes from the Latin compound verb *in-*

sultare, which means "to jump on" and is also the root word for *assault* and *assail.* It's a word that connotes aggressive physical abuse, an attack. Originally, it suggested leaping upon the prostrate body of a foe, which may be how its link to contemptuous action was forged.

Though "an insult to the brain" (a blow to the head, a metal shard through the skull, a stroke, a viral "attack") is a kind of assault, I am curious about the way *contempt* has found its way into the matter. Contempt was always part of the meaning of insult and now it is primary to the meaning. Certainly a virus is not acting contemptuously when it targets the brain; neither is the pavement nor steering wheel nor falling wrench nor clot of blood nor most other agents of "insult." But I think society at large, medical scientists, insurers, legislators, and the person on the street do feel a kind of contempt for the brain damaged with their comical way of walking, their odd patterns of speech or ways in which neurological damage is expressed, their apparent stupidity, their abnormality. The damage done to a brain seems to evoke disdain in those who observe it and shame or disgrace in those who experience it.

Poet Peter Davison has noticed the resonant irony of the phrase "an insult to the brain" and made use of it in his poem "The Obituary Writer." Thinking about the suicide of John Berryman, the heavily addicted poet whose long-expected death in 1972 followed years of public behavior symptomatic of brain damage, Davison writes that "his hullabaloos / of falling-down drunkenness were an insult to the brain." In this poem, toying with the meaning of the phrase, Davison suggests that Berryman's drinking may have been an insult to his brain, technically speaking, but that watching him was, for a friend, another kind of brain insult. He has grasped the fatuousness of the phrase as a medical term, its inherent judgment of contempt, and made use of it for its poetic ambiguity.

But I have become enamored of the idea that my brain has been insulted by a virus. I use it as motivation. There is a long tradition of avenging insults through duels or counter-insults, through litigation, through the public humiliation of the original insult. So I write. I avenge myself on an insult that was meant, it feels, to silence me by compromising my word-finding capacity, my ability to concentrate and remember, to spell or conceptualize, to express myself, to think.

The duel is fought over and over. I have developed certain habits that enable me to work—a team of seconds, to elaborate this metaphor of a duel. I must be willing to write slowly, to skip or leave blank spaces where I cannot find words that I seek, compose in fragments and without an overall ordering principle or imposed form. I explore and make discoveries in my writing now, never quite sure where I am going, but willing to let things ride and discover later how they all fit together. Every time I finish an essay or poem or piece of fiction, it feels as though I have faced down the insult.

IN HIS BOOK *Creating Mind,* Harvard neurobiologist John E. Dowling says "the cerebral cortex of the human brain, the seat of higher neural function—perception, memory, language, and intelligence—is far more developed than is the cerebral cortex of any other vertebrate." Our gray matter is what makes us human. Dowling goes on to say that "because of the added neural cells and cortical development in the human brain, new facets of mind emerge." Like the fractured facet of a gemstone or crystal, like a crack in the facet of a bone, a chipped facet of mind corrupts the whole, and this is what an insult to the brain does.

Though people long believed, with Aristotle, that the mind was located within the heart, the link between brain and mind is by now a basic fact of cognitive science. Like countless others, I am living proof of it. Indeed, it is by studying the behavior of brain-damaged patients like me that medical science first learned, for example, that the brain is modular, with specific areas responsible for specific functions, or that functions on one side of the body are controlled by areas on the opposite side of the brain. "The odd behavior of these patients," says Ramachandran, speaking of the brain-damaged, "can help us solve the mystery of how various parts of the brain create a useful representation of the external world and generate the illusion of 'self' that endures in space and time." Unfortunately, there is ample opportunity to observe this in action since, according to the Brain Injury Association, more than two million Americans suffer traumatic brain injury every year, a total that does not include damage by disease.

"Change the brain, change the person," says Restak in *The Modular Brain.* But how, exactly? No one has yet explained the way a brain produces what we think of as consciousness. How does the firing of electrical impulse across a synapse produce love, math, nightmare, theology, appetite? Stated more traditionally, how do brain and mind interact? Bookstore shelves are now filled with books, like Steven Pinker's brilliant 1997 study, *How the Mind Works,* which attempt to explain how a three-pound organ the consistency of Jell-O makes us see, think, feel, choose, and act. "The mind is not the brain," Pinker says, "but what the brain does."

And what the brain does, according to Pinker, "is information processing, or computation." We think we think with our brain. But in doing its job of creating consciousness, the brain actually relies upon a vast network of systems and is connected to everything—eyes, ears, skin, limbs, nerves. As Dowling so dourly puts it, our mental function, our mind—memory, feelings, emotions, awareness, understanding, creativity—"is an emergent property of brain function." In other words, "What we refer to as mind is a natural consequence of complex and higher neural processing."

The key word is *processing*. We actually think with our whole body. The brain, however, takes what is shipped to it, crunches the data, and sends back instructions. It converts; it generates results. Or, when damaged, does not. There is nothing wrong with my sensory receptors, for instance. I see quite well. I can hear and smell; my speech mechanisms (tongue, lips, nerves) are intact. My skin remains sensitive. But it's in putting things together that I fail. Messages get garbled, blocked, missed. There is, it sometimes seems, a lot of static when I try to think, and this is the gray area where nothing is clear any longer.

Neurons, the brain's nerve cells, are designed to process information. They "receive, integrate and transmit," as Dowling says, receiving input from dendrites and transmitting output along axons, sending messages to one another across chemical passages called *synapses*. When there are lesions like the ones that riddle my gray matter, processing is compromised. Not only that, certain cells have simply died and with them the receiving, integrating and transmitting functions they performed.

My mind does not make connections because, in essence, some of my brain's connectors have been broken or frayed. I simply have less to work with and it is no surprise that my IQ dropped measurably in the aftermath of my illness. Failing to make connections, on both the physical and metaphysical levels, is distressing. It is very difficult for me to free-associate; my stream of consciousness does not absorb runoff or feeder streams well, but rushes headlong instead. Mental activity that should follow a distinct pattern does not and, indeed, I experience my thought process as subject to random misfirings. I do not feel in control of my intelligence. Saying "Pass me the tracks" when I intended to say "Pass me the gravy" is a nifty example. Was it because *gravy* sounds like *grooves* which led to tracks or because my tendency to spill gravy leaves tracks on my clothes? A misfire, a glitch in the gray area that thought has become for me, and as a result my ability to express myself is compromised. My very nature seems to have altered.

I am also easily overloaded. I cannot read the menu or converse in a crowded, noisy restaurant. I get exhausted at Portland Trailblazers basketball games, with all the visual and aural imagery, all the manufactured commotion, so I stopped going nine years ago. My hands are scarred from burns and cuts that occurred when I tried to cook and converse at the same time. I cannot drive in traffic, especially in our standard transmission pickup truck. I cannot talk about, say, the fiction of Thomas Hardy while I drive; I need to be given directions in small doses rather than all at once, and need those directions to be given precisely at the time I must make the required turn. This is, as Restak explains, because driving and talking about Hardy, or driving and processing information about where to turn, are handled by different parts of the brain and my brain's parts have trouble working together.

I used to write accompanied by soft jazz, but now the least pattern of noises distracts me and shatters concentration. My entire writing process, in fact, has been transformed as I learned to work with my newly configured brain and its strange snags. I have become an avid note taker, a jotter of random thoughts that might or might not find their way together or amount to anything, a writer of bursts instead of steady work. A slight interruption—the movement of my cat across my window view, the call of a hawk, a spell of coughing—will not just make me lose my train of thought, it will leave me at the station for the rest of the day.

I have just finished reading a new book about Muhammad Ali, *King of the World,* by David Remnick. I anticipated identifying a bit with Ali, now suffering from Parkinson's disease, who shows so strikingly what brain damage can do, stripped as he is of so many of the functions—speech, movement, spontaneity—that once characterized him. But it was reading about Floyd Patterson that got me.

Patterson was a childhood hero of mine. Not only did we share a rare first name, we lived in neighboring towns—he was in Rockville Center, on Long Island, while I was five minutes away in Long Beach, just across the bridge. I was nine when he beat Archie Moore to take the heavyweight championship belt, almost twelve when he lost it to Ingemar Johansson and almost thirteen when he so memorably won it back. The image of Johansson's left leg quivering as he lay unconscious on the mat is one of those vivid memories that endures (because, apparently, it is stored in a different part of the brain than other, less momentous memories). That Floyd, like me, was small of stature in his world, was shy and vulnerable, and I was powerfully drawn to him.

During his sixty-four professional fights, his long amateur career, his many rounds of sparring to prepare for fights, Patterson absorbed a tremendous amount of damage to his brain. Now in his sixties, his ability to think is devastated. Testifying in court earlier this year in his capacity as head of the New

York State Athletic Commission, Patterson "generally seemed lost," according to Remnick. He could not remember the names of his fellow commissioners, his phone number or secretary's name or lawyer's name. He could not remember the year of his greatest fight, against Archie Moore, or "the most basic rules of boxing (the size of the ring, the number of rounds in a championship fight)." He kept responding to questions by saying, "It's hard to think when I'm tired."

Finally, admitting "I'm lost," he said, "Sometimes I can't even remember my wife's name, and I've been married thirty-two, thirty-three years." He added again that it was hard for him to think when he was tired. "Sometimes, I can't even remember my own name."

PEOPLE OFTEN ASK if I will ever "get better." In part, I think what they wonder about is whether the brain can heal itself. Will I be able, either suddenly or gradually, to think as I once did? Will I toss aside the cane, be free of symptoms, have all the functions governed by my brain restored to smooth service, rejoin the world of work and long-distance running? The question tends to catch me by surprise because I believe I have stopped asking it myself.

The conventional wisdom has long been that brains do not repair themselves. Other body tissue, other kinds of cells, are replaced after damage, but "When brain cells are lost because of injury or disease," Dowling wrote as recently as 1998, "they are not replaced." We have, he says, as many brain cells at age one as we will ever have. This has been a fundamental tenet of neuroscience, yet it has also long been clear that people do recover—fully or in part—from brain injury. Some stroke victims relearn to walk and talk; feeling returns in once-numb limbs. Children—especially children—recover and show no lasting ill effects from catastrophic injuries or coma-inducing bouts of meningitis.

So brain cells do not get replaced or repaired, but brain-damaged people occasionally do regain function. In a sense, then, the brain heals, but its cells do not.

In "Confronting Traumatic Brain Injury," Texas bioethicist William J. Winslade says, "Scientists still don't understand how the brain heals itself." He adds that although "Until recently, neuroscientists thought that much of the loss of capabilities due to brain damage was irreversible," patients recover spontaneously and rehabilitation programs "can restore cognitive and functional skills and emotional and experiential capacity, at least in part."

There are in general five theories about the way people might recover function lost to brain damage. One suggests that we do not need all of our brain because we only use a small part of it to function. Another is that some brain tissue can be made to take over functions lost to damage elsewhere. Connected to this is the idea that the brain has a backup mechanism in place allowing cells to take over like understudies. Rehabilitation can teach people new ways to perform some old tasks, bypassing the whole damaged area altogether. And finally, there is the theory that in time, and after the chemical shock of the original injury, things return to normal and we just get better.

It is probably true that, for me, a few of these healing phenomena have taken place. I have, for instance, gotten more adept at tying my shoes, taking a shower, driving for short periods. With careful preparation, I can appear in public to read from my work or attend a party. I have developed techniques to slow down my interactions with people or to incorporate my mistakes into a longer-term process of communications or composition. I may not be very good in spontaneous situations, but given time to craft my responses I can sometimes do well. But I still can't think.

A recent development promises to up the ante in the game of recovery from brain damage. The *New York Times* reported in October of 1998 that "Adult humans can generate new brain cells." A team at the Salk Institute for Biological Studies in La Jolla, California, observed new growth in cells of the hippocampus, which controls learning and memory in the brain. The team's leader, Dr. Fred Gage, expressed the usual cautions; more time is needed to "learn whether new cell creation can be put to work" and under what conditions. But the findings were deemed both "interesting" and "important."

There is only one sensible response to news like this. It has no personal meaning to me. Clinical use of the finding lies so far in the future as to be useless, even if regenerating cells could restore my lost functions. Best not to think about this sort of thing.

Because, in fact, the question of whether I will ever get better is meaningless. To continue looking outside for a cure, a "magic bullet," some combination of therapies and treatments and chemicals to restore what I have lost is to miss the point altogether. Certainly if a safe, effective way existed to resurrect dead cells, or generate replacements, and if this somehow guaranteed that I would flash back or flash forward to "be the person I was," it would be tempting to try.

But how would that be? Would the memories that have vanished reappear? Not likely. Would I be like the man, blind for decades, who had sight restored and could not handle the experience of vision, could not make sense of a world he could see? I am, in fact, who I am now. I have changed. I have learned to live and live richly as I am now. Slowed down, softer, more heedful of all that I see and hear and feel, more removed from the hubbub, more internal. I have made certain decisions, such as moving from the city to a remote rural hilltop in the middle of acres of forest, that have turned out to be good for my health and even my soul. I have gained the love of a woman who knew me before I got sick and likes me much better now. Certainly I want to be well. I miss being able to think clearly and sharply, to function in the world, to move with grace. I miss the feeling of coherence or integrity that comes with a functional brain. I feel old before my time.

In many important respects, then, I have already gotten better. I continue to learn new ways of living with a damaged brain. I continue to make progress, to avenge the insult, to see my way around the gray area. But no, I am not going to be the man I was. In this, I am hardly alone.

DENIS G. PELLI

Close Encounters: An Artist Shows That Size Affects Shape

FROM *SCIENCE*

Visual perception plays a crucial role in mental functioning. How we see helps determine what we think and what we do. Scientists can perform laboratory experiments to get at the mechanism of seeing, but Denis G. Pelli, a professor of psychology and neural science at New York University, has had the happy idea of enlisting the visual arts. Seizing on the unique properties of Chuck Close's remarkable block portraits, in which faces are built from distinct geometrical units, Pelli dismisses conventional notions of the ways in which size and shape affect how we perceive objects.

We easily recognize objects of all shapes and sizes, yet no one has any idea how we do it. It seems obvious that we must see shape in the same way regardless of size, otherwise we would recognize our friend or the letter "a" differently at each size or viewing distance.[1] Yet the block portraits by the artist Chuck Close vividly show the size dependence of shape perception. When viewing any of Close's 1987–1997 portraits at their actual size,[2] one can move forward and back, again and again, and the face, solid from afar, always collapses into flat marks when seen from near. The duality (solid from afar and flat from near) of these paintings shows that size affects the perception of shape, disproving the popular assumption that shape perception is size-independent. We have reproduced a recent Close block portrait, *Bill II* (1991), at one-third of its actual size to allow readers to experience the dramatic effect of size on what they perceive. Psychophysical measurements of observers viewing Close's block portraits reveal the importance of size, that the effect is visual (perception) not optical (physics), and that it involves a competition between the face and its constituent blocks to engage our perception of shape from shading.

Aristotle[3] noted that shape perception could be independent of size only for sizes that are neither so huge as to exceed our visual field (~135° visual angle subtended at the eye) nor so tiny as to exceed our visual acuity (0.1°). As visual acuity (the fineness of vision) is largely determined by optics, the assumption among vision scientists has been that observers identify the shape of the (blurry) retinal image independently of its size. This age-old assumption seems to be supported by both common experience and priming experiments (in which the speed and accuracy of target identification are unaffected by the size difference between target and prime images).[4]

Nearly all of Close's paintings are of heads produced by photographing the subject, drawing a grid onto the photo and a similar grid onto the canvas, and meticulously copying one square at a time from photo to canvas. Seen from a great distance, the portrait is a visually accurate reproduction of the photo. In his earliest paintings, Close copied details within each square so that the original grid is not visible in the result, but since 1973 he has usually filled each square of the grid on his canvas with content that is independent of the original photo. Close refers to each filled square as a "mark."

In 1973, the scientists Harmon and Julesz published block portraits of Abraham Lincoln that were important in vision science.[5] Their "critical-band" explanation of the block-portrait effect implicitly assumes size invariance and thus predicts that face recognition requires a certain number of marks per face independent of face size (but they only tested the effect at one small size). Close, too, showed his first block portrait (dot drawing) in 1973.[6] Unlike Close's recent work, the early block portraits by Close, and those by Harmon and Julesz, had small (<1 cm) blocks that viewers never approached closely enough to experience the full duality.[7] Instead, viewers experienced only a weak one-time effect, elegantly described by Harmon: "Viewed from close up, these 'block portraits' appear to be merely an assemblage of squares, . . . [but]

once a face is perceived it becomes difficult not to see it, as if some kind of perceptual hysteresis prevented the image from once again dissolving into an abstract pattern of squares."

There are well-known antecedents to block paintings such as the coarse "benday" screens used in Roy Lichtenstein's cartoons of the 1960s, Seurat's pointillist *Sunday Afternoon on the Island of La Grande Jatte* (1884–1886), and the ancient mosaics at Delos and Pompeii. However, none used coarse grids to render three-dimensional shape, so there is no duality. The mosaics and the pointillist grids are too fine to readily disintegrate into flat marks, and the benday screens are uniform, so that they are always flat at all distances. The only precedent for the duality of Close's recent heads may be the long-lost 4th century (B.C.E.) *skiagraphia* paintings by Apollodorus in which he achieved intermediate colors by juxtaposing large patches of unmixed colors that blended when viewed from a distance.[8]

The Chuck Close retrospective[2] exhibits scores of block portraits, half in color (like *Bill II)* and half in black and white, with a wide range of marks across the face (5 to 21), face size (2 to 200 cm), mark size (0.5 to 9 cm), and mark type. This variety allowed us to undertake a parametric investigation of the size dependence of shape perception. We used a psychophysical "nose test" to measure the transition from flat to solid in 33 Close portraits. While looking at the painting, the observer is asked to move forward and backward to find the viewing distance at which the nose emerges from the canvas. (The instructions emphasize the bridge of the nose to minimize the effect of the nostrils, which Close usually renders with detail exceeding that of the grid.) From

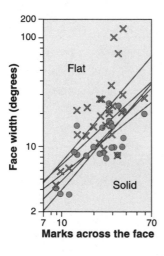

The critical face width of Close's portraits plotted against the number of marks across the face. The regression lines, one for each of the five observers (accounting for half of the variance), are plots of the results for judging nose emergence on each of the 33 gridded portraits from the Chuck Close retrospective (except the Keith/Six Drawings Series)[2]. X and o are raw data for two observers. Size independence predicts a vertical line. All five lines have log-log slopes close to 1 (mean 1.0, SD ± 0.2), showing that the perceived shape does depend on size.

afar, the nose (illuminated from the side) appears to stick out from the canvas. As the viewer approaches, the nose suddenly collapses, becoming a flat patch of unevenly colored skin. The transition is abrupt. The observer is asked to find the transition point, and the critical distance is measured from the observer's eye to the painting. Clearly, size does matter because the face is perceived differently at different sizes (viewing distances).

When viewing all possible block portraits there is a division into two domains: The face is seen as flat in one domain and solid in the other. We mapped the boundary dividing the two domains to reveal how size affects shape. For a given painting, the number of marks across the face is fixed. The critical face width in degrees is the angle subtended by the face at the critical viewing distance. We plotted critical face width against the number of marks across the face (see graph). The five solid lines represent the results from the viewing of 33 block portraits by five observers. There are obvious differences between observers, presumably because each must set his or her own internal criterion for nose emergence. If shape perception were size-independent, the plot would be a vertical line. Instead, the modest positive slopes indicate that people need more marks to see the nose stick out on larger faces. The five lines have slopes close to 1, demonstrating that critical face width is proportional to number of marks across the face (see graph). From the slope of the lines it is clear that the mark size (face width in degrees divided by number of marks across the face) is constant. So, it is the critical mark size (roughly 0.3°) that divides the flat from the solid domain. Portraits are seen as flat when marks are larger than 0.3°, and solid when marks are smaller than 0.3°, for faces of all sizes.

The type of mark used may account for much of the residual scatter of the points about the lines in the graph. Close has tried many types of mark, and they affect the critical distance at which the nose collapses. The *Keith/Six Drawings Series* (1979) of Close's portraits are all the same size, based on the same photograph and grid, but use very different marks (watercolor dots, fingerprints, ink stick scribbles, and white Conté crayon) and have different critical mark sizes, ranging from 0.2° to 0.7°.

But can the size effects of Close's paintings be explained by simple optics? Vision scientists, taking size independence for granted, have supposed that increasing the viewing distance reveals the face simply because it increases the blur, wiping out the grid. By taking off one's glasses (or putting on someone else's), one can blur the image (remove the grid) and the face is revealed, as viewers of Close's paintings often discover for themselves. But, whereas one could (at least in principle) walk far enough away from a block portrait to achieve the same blur as the wrong glasses provide at short distances, in fact,

Bill II (1991), a block portrait by Chuck Close, reduced to one-third of its actual size and cropped. Compare its appearance from near and far (>5m) or compare it with the tiny copy of *Bill II* (see page 108). Ignore the pupils, nostrils, and the line between the lips, which have much higher resolution than the 1.3-cm grid that represents the rest of the face. Below each letter in the eye chart is a number indicating its size (the observer's acuity) as a fraction of a mark (a filled square of the painting's grid). [Oil on canvas, 36×30″ (92.4×76.2 cm). *(Photograph by Bill Jacobson, courtesy of Pace Wildenstein. Copyright © by Chuck Close.)*

the size effects that are most salient in the Close exhibition occur over modest distances (<6 m) at which the eye's blur is only a fraction of a mark. Readers can try this for themselves by finding the points of nose emergence with and without their glasses, and comparing the appearance of *Bill II* and their visual

Chuck Close's *Maggie* (1996). Oil on canvas, 30x24" (76.2x61 cm). Reduced and cropped. This painting demonstrates that the perception of shape depends on the actual size of the image. Compare its appearance from far to near, as the solid face collapses into flat squares. *(Photograph by Ellen Page Wilson, courtesy of Pace Wildenstein. Copyright © by Chuck Close.)*

acuity (size of smallest readable letter) between the two conditions. Defocus (blur) reveals the nose only when it completely smears out the grid, at an acuity of about one mark. But when the observer simply increases viewing distance, without defocus, the grid is still apparent when the nose pops out, at an acuity of about a quarter mark.

To unequivocally conclude that perceived shape depends on the size per se of the retinal image, we tested a condition in which the retinal image changes only in size. We compared critical distances while looking through either a 1- or 2-mm pinhole; these artificial pupils are sufficiently small to make the eye diffraction-limited (the quality of the image depends only on pupil size).[9] The eye's blur with the 1-mm pupil is twice as big as, but otherwise identical to, its blur with the 2-mm pupil. (The retinal illuminances are equated by adding a 25% transmission neutral density filter to the 2-mm pupil.) If we perceived the retinal image's shape independently of its size, then the critical image size should double when we double the blur. In fact, we find that the critical image sizes are identical, showing that retinal image size, not blur, determines the perceived shape.

Observers must see marks both substantially larger and smaller than 0.3° to experience duality. To collapse reliably, a portrait composed of 0.4-cm marks (like Harmon and Julesz's Lincoln) must be viewed from less than 40 cm, closer than most viewers will come. For 2-cm marks, or larger, as in Close's recent 1987–97 portraits, that distance is at least 2 m, which most viewers cross as they approach. Some forms of camouflage, like a tiger's stripes, may break up the animal's shape only when seen from very near. Most perception textbooks show a spotted Dalmatian, initially lost in a background of spotty shadows, but which usually appears quickly and never goes away (like Harmon's description of seeing Lincoln with small blocks).[10] I find that the Dalmatian, like a block portrait, does break up reliably into mere flat spots when enlarged (or approached) to make the spot spacing exceed 0.3°.

Testing a wide range of sizes revealed that the division between seeing a block portrait as flat or solid occurs at a critical mark size of 0.3° (which is independent of the number of marks per face). This refutes the size invariance of shape perception and Harmon and Julesz's critical-band theory of the block-portrait effect. It seems that the blocks (or their edges)[11] compete with the face to capture the visual shape-from-shading process. The size and type of the marks determine their power of attraction. This competition is bottom-up, determined by the stimulus, not top-down, controlled by the observer. Close concedes that, painting at arm's length, even he cannot see the face unless he backs away.[2]

One might suppose that Close was a naïve artist, obsessed by grids, who innocently produced the coarsely gridded paintings that we use here to reveal the size dependence of shape perception. In fact, Close has devoted his career to studying just that: "The self portrait from 1967–68 is the first portrait head that I painted. . . . The idea was to make something that was so large that it could not be readily seen as a whole and force the viewer to scan the image in a Brobdingnagian way, as if they were Gulliver's Lilliputians crawling over the surface of the face, falling into a nostril and tripping over a mustache hair."[2] He was more thorough than his scientific colleagues; the size of the marks in his block portraits increased by 15% per year from 1973 (0.4 cm) to 1997 (9 cm). He made sure that exhibitions of his work would convey the idea, canceling a retrospective that could not provide long viewing distances. So credit Chuck Close with discovering this size-dependent breakdown of our ability to extract shape from shading, well within the bounds of our visual field and acuity.

References and Notes

1. In 1886, Ernst Mach wrote, "Some forty years ago, in a society of physicists and physiologists, I proposed for discussion the question, why geometrically similar figures were also optically [visually] similar. I remember quite well the attitude taken with regard to this question, which was accounted not only superfluous, but even ludicrous" [*The Analysis of Sensations,* translated by C. M. Williams and S. Waterlow (Routledge/Thoemes, London, 1996), p. 109].

2. Chuck Close exhibition organized by Robert Storr, Museum of Modern Art, New York City, 26 February to 26 May 1998. The exhibition was at the Hayward Gallery, London, 22 July to 19 September, 1999. All relevant paintings, except *Maggie,* are reproduced in the catalog [R. Storr, *Chuck Close* (Museum of Modern Art, New York, 1998)]. The Close quote is from recorded narration provided at the exhibition.

3. "Beauty depends on size and order; hence an extremely minute creature could not be beautiful, for our vision becomes blurred as it approaches the point of imperceptibility, nor could an utterly huge creature be beautiful, for, unable to take it in all at once, the viewer finds that its unity and wholeness have escaped his field of vision" [*Aristotle's Poetics*, translated by J. Hutton (Norton, New York, 1982), chapter 7].

4. E. E. Cooper, I. Biederman, J. E. Hummel, *Can. J. Psychol.* **46,** 191 (1992). Unlike identification, judging whether one has seen a particular object before is size-specific. For discussion of the size question, see <www.visionscience.com/mail/cvnet/1998/0307.html>.

5. L. D. Harmon and B. Julesz, *Science* **180,** 1194 (1973); L. D. Harmon, *Sci. Am.* **229** (no. 5), 71 (1973); in O. J. Grusser and R. K. Klinke (Eds.), *Pattern Recognition in Biological and Technical Systems* (Springer-Verlag, New York, 1971), pp. 196–219.

6. *Keith/1,280,* exhibited at the Bykert Gallery, New York City, 20 October to 15 November 1973.

7. Harmon (1971) includes a face with 0.4 cm squares. The blocks are 0.4 cm in Harmon and Julesz's (1973) Lincoln (cover of *Science*), 0.8 cm in Harmon's (1973) George Washington (cover of *Scientific American*), and 1.0 cm in his Mona Lisa. Close's *Keith/1,280* (1973) is made up of dots on a 0.4-cm grid.

8. E. C. Keuls [*Plato and Greek Painting* (Brill, Leiden, Netherlands, 1978)], on the basis of texts by Plato and Aristotle.

9. F. W. Campbell and R. W. Gubisch, *J. Physiol.* **186,** 558 (1966).

10. Dalmatian photo by R. C. James, in R. L. Gregory, *The Intelligent Eye* (McGraw-Hill, New York, 1973), p. 14.

11. M. C. Morrone, D. C. Burr, J. Ross, *Nature* **305,** 226 (1983).

Douglas R. Hofstadter

Analogy as the Core of Cognition

Few are more stimulating on the subject of how we think than Douglas Hofstadter, who has brought his fertile imagination to a breathtakingly diverse array of intellectual enterprises—from mathematical puzzles to translating Pushkin. In this lively and far-ranging essay, the author of Gödel, Escher, Bach *posits that the making of analogies does not grow out of human reasoning ability, but, more fundamentally, stands at the very basis of thought and makes human reasoning possible.*

Grand Prelude and Mild Disclaimer

ONCE UPON A TIME, I was invited to speak at an analogy workshop in the legendary city of Sofia in the far-off land of Bulgaria. Having accepted but wavering as to what to say, I finally chose to eschew technicalities and instead to convey a personal perspective on the importance and centrality of analogy-making in cognition. One way I could suggest this perspective is to re-chant a refrain that I've chanted quite oft in the past, to wit:

One should not think of analogy-making as a special variety of *reasoning* (as in the dull and uninspiring phrase "analogical reasoning and problem-solving", a long-standing cliché in the cognitive-science world), for that is to do analogy a terrible disservice. After all, reasoning and problem-solving have (at least I dearly hope!) been at long last recognized as lying far indeed from the core of human thought. If analogy were merely a special variety of something that in itself lies way out on the peripheries, then it would be but an itty-bitty blip in the broad blue sky of cognition. To me, however, analogy is anything but a bitty blip—rather, it's the very blue that fills the whole sky of cognition—analogy is *everything*, or very nearly so, in my view.

End of oft-chanted refrain. If you don't like it, you won't like what follows.

The thrust of my article is to persuade readers of this unorthodox viewpoint, or failing that, at least to give them a strong whiff of it. In that sense, then, my article shares with Richard Dawkins' eye-opening book *The Selfish Gene* [Dawkins 1976] the quality of trying to make a scientific contribution mostly by suggesting to readers a shift of viewpoint—a new take on familiar phenomena. For Dawkins, the shift was to turn causality on its head, so that the old quip "a chicken is an egg's way of making another egg" might be taken not as a joke but quite seriously. In my case, the shift is to suggest that every concept we have is essentially nothing but a tightly packaged bundle of analogies, and to suggest that all we do when we think is to move fluidly from concept to concept—in other words, to leap from one analogy-bundle to another—and to suggest, lastly, that such concept-to-concept leaps are themselves made via analogical connection, to boot.

This viewpoint may be overly ambitious, and may even—horrors—be somewhat wrong, but I have observed that many good ideas start out by claiming too much territory for themselves, and eventually, when they have received their fair share of attention and respect, the air clears and it emerges that, though still grand, they are not quite so grand and all-encompassing as their proponents first thought. But that's all right. As for me, I just hope that my view finds a few sympathetic readers. That would be a fine start.

Two Riddles

WE BEGIN WITH A COUPLE of simple queries about familiar phenomena: "Why do babies not remember events that happen to them?" and "Why does each new year seem to pass faster than the one before?"

I wouldn't swear that I have the final answer to either one of these queries, but I do have a hunch, and I will here speculate on the basis of that hunch. And thus: the answer to both is basically the same, I would argue, and it has to do with the relentless, lifelong process of *chunking*—taking "small" concepts and putting them together into bigger and bigger ones, thus recursively building up a giant repertoire of concepts in the mind.

How, then, might chunking provide the clue to these riddles? Well, babies' concepts are simply *too small*. They have no way of framing entire events whatsoever in terms of their novice concepts. It is as if babies were looking at life through a randomly drifting keyhole, and at each moment could make out only the most local aspects of scenes before them. It would be hopeless to try to figure out how a whole room is organized, for instance, given just a keyhole view, even a randomly drifting keyhole view.

Or, to trot out another analogy, life is like a chess game, and babies are like beginners looking at a complex scene on a board, not having the faintest idea how to organize it into higher-level structures. As has been well known for decades, experienced chess players chunk the setup of pieces on the board nearly instantaneously into small dynamic groupings defined by their strategic meanings, and thanks to this automatic, intuitive chunking, they can make good moves nearly instantaneously and also can remember complex chess situations for very long times. Much the same holds for bridge players, who effortlessly remember every bid and every play in a game, and months later can still recite entire games at the drop of a hat.

All of this is due to chunking, and I speculate that babies are to life as novice players are to the games they are learning—they simply lack the experience that allows understanding (or even perceiving) of large structures, and so nothing above a rather low level of abstraction gets perceived at all, let alone remembered in later years. As one grows older, however, one's chunks grow in size and in number, and consequently one automatically starts to perceive and to frame ever larger events and constellations of events; by the time one is nearing one's teen years, complex fragments from life's stream are routinely stored as high-level wholes—and chunks just keep on accreting and becoming more numerous as one lives. Events that a baby or young child could not have possibly perceived as such—events that stretch out over many minutes, hours, days, or even weeks—are effortlessly perceived and stored away as single structures with much internal detail (varying amounts of which can be pulled up and contemplated in retrospect, depending on context). Babies do not have large chunks and simply cannot put things together coherently. Claims by some people that they remember complex events from when they were but a few months old (some even claim to remember being born!) strike me as nothing more than highly deluded wishful thinking.

So much for question number one. As for number two, the answer, or so I would claim, is very similar. The more we live, the larger our repertoire of concepts becomes, which allows us to gobble up ever larger coherent stretches of life in single mental chunks. As we start seeing life's patterns on higher and higher levels, the lower levels nearly vanish from our perception. This effectively means that seconds, once so salient to our baby selves, nearly vanish from sight, and then minutes go the way of seconds, and soon so do hours, and then days, and then weeks . . .

"Boy, this year sure went by fast!" is so tempting to say because each year is perceived in terms of chunks at a higher, grander, larger level than any year preceding it, and therefore *each passing year contains fewer top-level chunks* than any year preceding it, and so, psychologically, each year seems sparser than any of its predecessors. One might, somewhat facetiously, symbolize the ever-rapider passage of time by citing the famous harmonic series:

$$1 + \tfrac{1}{2} + \tfrac{1}{3} + \tfrac{1}{4} + \tfrac{1}{5} + \tfrac{1}{6} + \tfrac{1}{7} + \tfrac{1}{8} \, . \, . \, .$$

by which I mean to suggest that one's nth year feels subjectively n times as short as one's first year, or $n/5$ times as short as one's fifth year, and so on. Thus when one is an adult, the years seem to go by about at roughly a constant rate, because—for instance—$(\tfrac{1}{35})/(\tfrac{1}{36})$ is very nearly 1. Nonetheless, according to this theory, year 70 would still shoot by twice as fast as year 35 did, and seven times as fast as year 10 did.

But the exact numerical values shown above are not what matter; I just put them in for entertainment value. The more central and more serious idea is simply that relentless mental chunking makes life seem to pass ever faster as one ages, and there is nothing one can do about it. So much for our two riddles.

Analogy, Abstract Categories, and High-level Perception

BEFORE I GO ANY FURTHER, I would like to relate all this to analogy, for to some, the connection may seem tenuous, if not nonexistent. And yet to me, by contrast, analogy does not just lurk darkly here, but is right up there, front and center. I begin with the mundane observation that vision takes an input of millions of retinal dots and gives an output of concepts—often words or phrases, such as "duck," "Victorian house," "funky chair," "Joyce Carol Oates hairdo," or "looks sort of like President Eisenhower." The (visual) perceptual process, in other words, can be thought of as the triggering of mental categories—often standard lexical items—by scenes. Of course high-level perception can take place through other sensory modalities: we can hear a low rumbling noise and say "helicopter," can sniff something and remark "doctor's office," can taste something and find the words "okra curry" jumping to our tongue, and so on.

In fact, I should stress that the upper echelons of high-level perception totally transcend the normal flavor of the word "perception," for at the highest levels, input modality plays essentially no role. Let me explain. Suppose I read a newspaper article about the violent expulsion of one group of people by another group from some geographical region, and the phrase "ethnic cleansing," nowhere present in the article, pops into my head. What has happened here is a quintessential example of high-level perception—but what was the input medium? Someone might say it was *vision*, since I used my eyes to read the newspaper. But really, was I perceiving ethnic cleansing *visually*? Hardly. Indeed, I might have heard the newspaper article read aloud to me and had the same exact thought pop to mind. Would that mean that I had *aurally* perceived ethnic cleansing? Or else I might be blind and have read the article in braille—in other words, with my fingertips, not my eyes or ears. Would that mean that I had *tactilely* perceived ethnic cleansing? The suggestion is absurd.

The sensory input modality of a complex story is totally irrelevant; all that matters is how it jointly activates a host of interrelated concepts, in such a way that further concepts (e.g., "ethnic cleansing") are automatically accessed and brought up to center stage. Thus "high-level perception" is a kind of misnomer when it reaches the most abstract levels, but I don't know what else to call it, because I see no sharp line separating it from cases of recognizing "French impressionism" in a piece of music heard on the radio or thinking "Art Deco" when looking at a typeface in an advertisement.

The triggering of prior mental categories by some kind of input—whether sensory or more abstract—is, I insist, an act of analogy-making. Why is this? Because whenever a set of incoming stimuli activates one or more mental categories, some amount of slippage must occur (no instance of a category ever being precisely identical to a prior instance). Categories are quintessentially fluid entities; they adapt to a set of incoming stimuli and try to align themselves with it. The process of inexact matching between prior categories and new things being perceived (whether those "things" are physical objects or bite-size events or grand sagas) is analogy-making *par excellence*. How could anyone deny this? After all, it is the mental mapping onto each other of two entities—one old and sound asleep in the recesses of long-term memory, the other new and gaily dancing on the mind's center stage—that in fact differ from each other in a myriad of ways.

The Mental Lexicon: A Vast Storehouse of Triggerable Analogies

WE HUMANS BEGIN LIFE as rather austere analogy-makers—our set of categories is terribly sparse, and each category itself is hardly well-honed. Categories grow sharper and sharper and ever more flexible and subtle as we age, and of course fantastically more numerous. Many of our categories, though by no means all, are named by words or standard phrases shared with other people, and for the time being I will concentrate on those categories— categories that are named by so-called "lexical items." The public labels of such categories—the lexical items themselves—come in many grades, ranging more or less as follows:

- simple words: *chair, clock, cork, cannon, crash, clown, clue, cloak, climber . . .*
- compound words: *armchair, alarm clock, corkscrew, cannonball, skyscraper, station wagon, sexpot, salad dressing, schoolbus, jukebox, picket line, horror movie, wheeler-dealer . . .*
- short phrases: *musical chairs, out of order, Christmas tree ornament, nonprofit organization, business hours, foregone conclusion, rush-hour traffic, country-Western music, welcome home, tell me about it, give me a break, and his lovely wife, second rate, swallow your pride . . .*
- longer phrases: *stranded on a desert island; damned if you do, damned if you don't; praise the Lord and pass the ammunition; not in the foreseeable future; to the best of my knowledge; and they lived happily ever after; if it were up to me; haven't seen her since she was knee-high to a grasshopper; you could have knocked me over with a feather; thank you for not smoking; handed to him on a silver platter . . .*

Such lists go on and on virtually forever, and yet the amazing fact is that few people have any inkling of the vastness of their mental lexicons (I owe a major debt here to Joe Becker—see [Becker 1975]). To be sure, most adults use their vast mental lexicons with great virtuosity, but they have stunningly little explicit awareness of what they are doing.

It was Roger Schank, I believe, who pointed out that we often use *proverbs* as what I would call "situation labels," by which I mean that when we perceive a situation, what often springs to mind, totally unbidden, is some proverb tucked away in our unconscious, and if we are talking to someone, we will quote that proverb, and our listener will in all likelihood understand very clearly how the proverb "fits" the situation—in other words, will effortlessly make the mapping (the *analogy*, to stress what it is that we are talking about here) between the phrase's meaning and the situation. Thus the following kinds of phrases can easily be used as situation labels:

> That's the pot calling the kettle black if *I* ever saw it!
> It just went in one ear and out the other . . .
> Speak of the devil!
> When the cat's away the mice will play!

The Common Core Behind a Lexical Item

I NOW MAKE AN OBSERVATION that, though banal and obvious, needs to be made explicitly nonetheless—namely, things "out there" (objects, situations, whatever) that are labeled by the same lexical item have something, some core, in common; also whatever it is that those things "out there" share is shared with the abstract mental structure that lurks behind the label used for them. Getting to the core of things is, after all, what categories are for. In fact, I would go somewhat further and claim that getting to the core of things is what thinking itself is for—thus once again placing high-level perception front and center in the definition of cognition.

The noun "shadow" offers a good example of the complexity and subtlety of structure that lurks behind not just *some* lexical items, but behind every single one. Note, first of all, the subtle difference between "shadow" and "shade": we do not speak of cattle seeking *shadow* on a hot day, but *shade*. Many languages do not make this distinction, and thus they offer their native speakers a set of categories that is tuned slightly differently.

In many parts of the world, there are arid zones that lie just to the east of mountain ranges (e.g., the desert in Oregon just to the east of the Cascade mountains); these regions are standardly referred to as the mountain chain's "rain shadow."

What does one call the roughly circular patch of green seen underneath a tree after a snowfall? It could clearly be called a "snow shadow"—the region where snow failed to fall, having been blocked by an object.

A young woman who aspires to join her high-school swimming team, but whose mother was an Olympic swimmer, can be said to be "in the shadow of her mother." In fact, if she joins the team and competes, she might even be said to be "swimming in the shadow of her mother." And if she performs less well than her mother did, she will be said to be "overshadowed" by her mother.

One might say about a man who has had a bout with cancer but has recovered and is now feeling more secure about his health, "He is finally feeling more or less out of the shadow of his cancer." Along similar lines, many countries in Europe have recovered, to a large extent, from the ravages of World War II, but some might still be said to lie "in the shadow of World War II."

Another type of shadow cast by World War II (or by any war) lies in the skewed population distribution of any decimated group; that is, one imagines the human population as constituting a kind of flow of myriad tiny entities (individual people) down through the years (like that of photons or snowflakes through space), but long after the war's end, there are certain "regions" of humanity (e.g., certain ethnic groups) where the flow of births has been greatly reduced, much as if by an "obstacle" (namely, the millions of deaths in prior generations, whose effect continues to reverberate for many decades before gradually fading away, as a group's population replenishes itself).

There is of course no sharp line between cases where a word like "shadow" is used conventionally and cases where it is used in a novel manner; although "rain shadow" is something of a standard phrase, "snow shadow" (even though it is far easier to see) is less common. And notions like that of "population shadow" mentioned at the end are probably novel to most readers of this article, even though a closely related notion like "in the shadow of the war" is probably not new.

In short, the domain of the word "shadow" is a blurry region in semantic space, as is any human category, and—here I hark back to my initial refrain— that blur is due to the subtleties of mapping situations onto other situations— due, in other words, to the human facility of making analogies. The point is, a concept is a package of analogies.

Complex Lexical Items as Names of Complex Categories

OVER THE NEXT FEW PAGES I will present a potpourri of mental categories (via proxies—namely, their English-language lexical-item representations); I invite you to think, as you consider each item, just what it is that very different exemplars of the category in question tend to have in common. Thus:

- dog
- backlog
- probably
- probab-*lee!*

I interrupt the list momentarily to comment on the last two entries above, which of course are not nouns. (Who says nouns are the only mental categories? Obviously, verbs represent categories as well—but the same holds true, no less, for adjectives, adverbs, and so forth.) Some situations call forth the word "probably"; most do not. To some situations, the concept behind the word "probably" simply *fits*, while to most, it does not fit. We learn how to use the word "probably" over the course of years in childhood, until it becomes so ingrained that it never crosses our mind that "probably" is the name that English speakers give to a certain category of situations; it simply is *evoked* effortlessly and rapidly by those situations, and it is uttered without any conscious thought as to how it applies. It just "seems right" or "sounds right."

What, then about the word below it: "probab-*lee*"? This, too, is a lexical item in the minds of most native speakers of contemporary American English—perhaps not often used, perhaps more commonly heard than uttered by readers of this article, but nonetheless, we native speakers of American English all relate to hearing the word "probably" accented on its final rather than its initial syllable, and we all somehow realize the connotations hidden therein, though they may be terribly hard to articulate. I won't try to articulate them myself, but I would merely point out that this phonetic variant of the word "probably" fits only certain situations and not others (where the "situation" includes, needless to say, not just what is being talked about but also the mood of the speaker, *and* the speaker's assessment of the mood of the listener as well). Example: "Are our stupid leaders ever going to learn their lesson?" "Who knows? Maybe they're doomed to keep on repeating the mistakes of the past." "Mmm . . . Probab-*lee* . . ."

My point, with all the phrases cited above, is to bring to your conscious awareness the fact that there are certain situations that one could call "*probab-lee!* situations," no less than there are certain situations that are "*musical chairs situations*" or "*speak of the devil situations.*" In short, lexical items can be very abstract categories evoked by special classes of situations and not by others. This applies to adjectives, adverbs, prepositions, interjections, short and long phrases, and so on. Thus let me continue my list.

- Come *on*!
- Go for it!
- It's about time!
- Well, excuuuuuuuuuuse me!
- Let's not stand on ceremony!
- without batting an eyelash
- ain't

Lest the lowest item above seem puzzling, let me point out that the notorious contraction "ain't," although it is in a certain sense ungrammatical and improper, is nonetheless used very precisely, like pinpoint bombing, by politicians, reporters, university presidents, and the like, who carefully and deliberately insert it into their speech at well-timed moments when they know their audience almost expects it—it fits the context perfectly. For example, a general trying to justify a bombing raid might say, in describing the series of deadly skirmishes that provoked it, "I'm sorry, but a Sunday picnic it just *ain't.*" This is just one of many types of "ain't" situations. We native speakers know them when we hear them, and we likewise have a keen ear for *improper* uses of the word "ain't" by educated people, even if we ain't capable of putting our finger on what makes them inappropriate. (Curiously enough, shortly after drafting this paragraph, I came across an article in the *New York Times* about the failure of a test missile to hit its target, and a perfectly straight photo caption started out, "Two out of four goals ain't bad . . ." As I said above, even the most highly placed sources will use this "ungrammatical" word without batting an eyelash.)

"Suggestions" Imparted on the Soccer Field

As a non-native speaker of Italian watching the 1998 Soccer World Cup on Italian television, I was struck by the repeated occurrence of a certain term in the rapidfire speech of all the commentators: the word *suggerimento* (literally, "suggestion"). They kept on describing players as having given *suggerimenti* to other players. It was clear from the start that a *suggerimento* was not a verbal piece of advice (a suggestion in the most literal sense), but rather some kind of pass from one player to another as they advanced downfield. But what kind of pass was it exactly? By no means were all passes called *suggerimenti;* this term was clearly reserved for events that seemed to have some kind of scoring potential to them, as if one player was wordlessly saying to another, "Here now—take this and *go for it!*"

But how does a sports announcer unconsciously and effortlessly distinguish *this* kind of pass from other passes that in many ways look terribly similar? When is this kind of nonverbal "suggestion" being given by one player to another? I sensed that this must be a subtle judgment call, that there's no black-and-white line separating *suggerimenti* from mere *passaggi*, but that nonetheless there is a kind of core to the concept of *suggerimento* that all Italian announcers and keen Italian observers of soccer would agree on, and that there are fringes of the category, where some people might feel the word applied and others would not. Such blurriness is the case, of course, with every mental category, ranging from "chair" to "wheeler-dealer" to "pot calling the kettle black," but since *suggerimento* was not in my native language and thus I had been forced to grapple with it explicitly and consciously, it was an excellent example of the view of lexical items that I am herein trying to impart to my readers.

Polysemy and the Nonspherical Shapes of Concepts

IT WOULD BE NAÏVE to imagine that each lexical item defines a perfectly "spherical" region in conceptual space, as pristine as an atomic nucleus surrounded by a spherical electron cloud whose density gradually attentuates with increasing distance from the core. Although the single-nucleus spherical-cloud image has some truth to it, a more accurate image of what lies behind a typical lexical item might be that of a molecule with two, three, or more nuclei that share an irregularly shaped electron cloud.

Suggerimento provides a perfect example of such a molecule, with one of its constituent atoms being the notion of a verbal piece of advice, another the notion of prompting on a theater stage, yet a third being the notion of a certain type of downfield soccer pass, and so forth. There is something in common, of course, that these all share, but they are nonetheless distinguishable regions in conceptual space.

Often native speakers of a language have a hard time realizing that two notions labeled identically in their language are seen as highly distinct concepts by speakers of other languages. Thus, native speakers of English feel the verb "to know" as a monolithic concept, and are sometimes surprised to find out that in other languages, one verb is used for knowing *facts*, a different verb for knowing *people*, and there may even be a third verb for knowing *how to do things*. When they are first told this, they are able to see the distinction, although it may seem highly finicky and pointless; with practice, however, they build up more refined categories until a moment may come when what once seemed an unnatural and gratuitous division of mental space now seems to offer a useful contrast between rather distinct notions. And conversely, speakers of a language where all three of these notions are represented by distinct lexical items may find it revelatory, fascinating, and perhaps even elegant to see how they are all subsumed under one umbrella-word in English.

My main point in bringing this up is simply to make explicit the fact that words and concepts are far from being regularly shaped convex regions in mental space; polysemy (the possession of multiple meanings) and metaphor make the regions complex and idiosyncratic. The simplest concepts are like isolated islands in a sea; the next-simplest are like pairs of islands joined by a narrow isthmus; then there are trios with two or three isthmuses having various widths; and so on. *Caveat:* When I say "simplest concepts," I do not mean those concepts that we pick up earliest in life, but in fact quite the contrary. After all, the majority of concepts planted in earliest childhood grow and grow over a lifetime and turn into the most frequently encountered concepts, whose elaborate ramifications and tendrils constitute the highest degree of twistiness! What I mean by "simplest concept" is merely "concept with maximally simple shape"; such a "simple" concept would most likely owe its simplicity precisely to its low frequency, and thus would seem like a sophisticated adult concept, such as "photosynthesis" or "hyperbola."

Conceptual Families and Lexical Rivalry

WALKING DOWN THE CORRIDORS of a building in Italy in which I have worked over several summers, I have been faced innumerable times with an interesting problem in high-level perception that has to be solved in real time—in a couple of seconds at most, usually. That is, how do I greet each person whom I recognize as we approach each other in the hall, and then pass? Here are five sample levels of greeting (there are dozens more, needless to say):

- *Buongiorno!* ("Hello!" or perhaps "Morning.")
- *Salve!* ("Howdy!" or perhaps "How are you.")
- *Buondì!* (Perhaps "Top o' the mornin'!" or "How ya doin'?")
- *Ciao!* ("Hi!" or "Hi there!")
- *Come stai?* ("How are you doing?" or perhaps "What's up?")

Each of them conveys a particular level of mutual acquaintance and a particular position along the formality/informality spectrum. And of course it frequently happens that I recognize someone but can't even remember how often I've met them before (let alone remember what their name is or what their role is), and so I have to make a decision that somehow will allow me to cover at least two different levels of friendliness (since I'm really not sure how friendly we are!). The choice is incredibly subtle and depends on dozens if not hundreds of variables, all unconsciously felt and all slightly contributing to a "vote" among my neurons, which then allow just one of these terms (or some other term) to come bubbling up out of my dormant Italian mental lexicon.

Consider the following spectrum of phrases all having in a certain sense "the same meaning," but ranging from very vulgar to somewhat incensed to quite restrained to utterly bland:

- He didn't give a flying f***.
- He didn't give a good God damn.
- He didn't give a tinker's damn.
- He didn't give a damn.
- He didn't give a darn.
- He didn't give a hoot.
- He didn't care at all.
- He didn't mind.
- He was indifferent.

For many native speakers, there are situations that correspond to each of these levels of intensity. To be sure, some speakers might be loath to utter certain of these phrases, but true native-level mastery nonetheless entails a keen awareness of when each of them might be called for in, say, a movie, or simply coming out of the mouth of someone *else*. After all, a large part of native mastery of a language is deeply knowing how *other* people use the language, regardless of whether one oneself uses certain phrases. And thus, to reiterate our theme, there are *"He didn't give a good God damn* situations" and there are situations of a very different sort, which could be called *"He didn't care at all* situations," and so forth. Each of the above expressions, then, can be thought of as *the name of a particular type of situation,* but since these categories are much closer to each other than just randomly chosen categories, they constitute potential rivalries that may take place during the ultra-fast high-level perceptual act that underlies speech.

Lexical Blends as a Window onto the Mind

LEXICAL BLENDS, which are astonishingly common though very seldom noticed by speakers or by listeners, reveal precisely this type of unconscious competition among close relatives in the mental lexicon. A lexical blend occurs when a situation evokes two or more lexical items at once and fragments of the various evoked competitors wind up getting magically, sometimes seamlessly, spliced together into the vocalized output stream (see, for example, [Hofstadter & Moser 1989]). Occasionally the speaker catches such an error on its way out and corrects it, though just as often it goes totally unheard by all parties. Thus people make blends of the following sorts:

- word-level blends: *mop/broom* → *brop*
- phrase-level blends: *easygoing/happy-go-lucky* → *easy-go-lucky*
- sentence-level blends: *We'll leave no stone unturned/We'll pull out all the stops* → *We'll pull no stops unturned.*

Blends reveal how much goes on beneath the surface as our brains try to figure out how to label simpler and more complex situations. In a way, what is amazing is that blends are not more common. Somehow, through some kind of cerebral magic, speakers light most of the time upon just one lexical label despite the existence of many potential ones, rather than coming out with a mishmosh of several—much as when a good pianist plays the piano, it is very seldom that two keys are struck at once, even though it might seem, *a priori,* that striking two neighboring keys at once ought to happen very often.

A Lexical Item as One Side of a Perceptual Analogy

AT THE RISK of boring some readers, I shall now continue with my rather arbitrary sampler of lexical items, just to drive the point home that *every lexical item that we possess is a mental category,* and hence, restating what I earlier claimed, *every lexical item, when used in speech (whether received or transmitted), constitutes one side of an analogy being made in real time in the speaker's/listener's mind.* I thus urge readers to try on for size the mindset that equates a lexical item with the "name" of a certain blurry set of situations centered on some core. Though this sounds quite orthodox for nouns, it is less so for verbs, and when applied to many of the following linguistic expressions, it is highly unorthodox:

- slippery slope
- safety net
- shades of . . .
- Been there, done that.
- For*get* it!
- It was touch-and-go.
- take a turn for the worse
- Be my guest!
- Make my day!
- Fancy that!
- Put your money where your mouth is!
- I mean, . . .
- Well, . . .
- Don't tell me that . . .
- It's fine to [do X] and all, but . . .
- kind of [+ adj.]
- when it comes to the crunch . . .
- You can't have it both ways!
- . . . *that's* for sure!
- the flip side [of the coin] is . . .
- You had to be there.
- It's high time that . . .
- Whatever!

Consider the teen-ager's favorite rejoinder, "Whatever!" If one were to try to capture its meaning—its range of applicability—one might paraphrase it somewhat along these lines: "You think such and so, and I disagree, but let's just agree to disagree and move on . . ." It takes a good number of years before one has acquired the various pieces of cognitive equipment that underpin the proper usage of such a phrase (which again ties in with the fact that one cannot remember events from one's babyhood).

High-Level Mental Chunks That Lack Labels

ALTHOUGH LONG STOCK PHRASES like "Put your money where your mouth is!" might seem to stretch the notion of mental chunking to the limit, that's hardly the case. Indeed, such phrases lie closer to the beginning than to the end of the story, for each one of us also remembers many thousands of events in our personal lives that are so large and so idiosyncratic that no one has ever given them a name and no one ever will, and yet they nonetheless are sharp memories and are revealed for the mental categories they are by the fact that they are summoned up cleanly and clearly by certain situations that take place later, often many years later. Thus take this one sample mental chunk, from my own personal, usually dormant repertoire:

> that time I spent an hour or two hoping that my old friend Robert, whom I hadn't seen in two years but who was supposed to arrive from Germany by train sometime during that summer day in the little Danish fishing village of Frederikssund (which in a series of letters he and I had mutually picked out on maps, and in which I had just arrived early that morning after driving all night from Stockholm), might spot me as I lurked way out at the furthest tip of the very long pier, rather than merely bumping into me at random as we both walked around exploring the stores and streets and parks of this unknown hamlet

As its length suggests, this is a very detailed personal memory from many years ago (and indeed, I have merely sketched it for readers here—I could write pages about it), and might at first seem to be nothing at all like a *mental category*. And yet, how else can one explain the fact that the image of myself standing at pier's end tingling with unrealistic hope jumped instantly to mind some fifteen years later as I was idly seeking to rearrange the eight letters in the last name of Janet Kolodner, a new acquaintance, in such a way that they would spell a genuine English word? Without success, I had tried dozens of fairly "obvious" pathways, such as "rendlook," "leodronk," and "ondorkle," when out of the blue it occurred to me that the initial consonant cluster "kn," with its cleverly silent "k," might be the key to success, and I started excitedly trying this "brilliant idea." However, after exploring this strategy for a while, I realized, to my chagrin, that no matter how lovely it would be if the silent "k" were to yield a solution, the probabilities for such a clever coup were rapidly diminishing. And at the precise instant that this realization hit, the Frederikssund-pier image came swooshing up out of memory, an image to which I had devoted not even a split second of thought for many years.

There was, of course, a perfectly logical reason behind this sudden resurfacing—namely, a strong and rich analogy in which the mundane idea of merely walking around the fishing village mapped onto the mundane exploration of "rendlook" and cousins, in which the "romantic" idea of lingering way out at the tip of the pier mapped onto the "romantic" hope for an anagram beginning with the tricky "kn" cluster, and in which the growing recognition of the likelihood of failure of the more unlikely, more "romantic" strategies was the common core that bound the two otherwise remote events together.

The Central Cognitive Loop

ABSTRACT REMINDINGS of this sort have been noted here and there in the cognitive-science literature, and some attempts have been made to explain them (e.g., Roger Schank's *Dynamic Memory* [Schank 1982]), but their starring role in the phenomenon of cognition has not, to my knowledge, been claimed. It is my purpose to stake that claim.

To make the claim more explicit, I must posit that such a large-scale memory chunk can be thought of as being stored in long-term memory as a "node"—that is, something that can be retrieved as a relatively discrete and separable whole, or to put it metaphorically, something that can be pulled like a fish out of the deep, dark brine of dormant memory. Once this "fish" has been pulled out, it is thrown in the "bucket" of short-term memory (often called "working memory"), where it is available for scrutiny.

Scrutiny consists in the act of "unpacking" the node to some degree, which means that inside it are found other nodes linked together by some fabric of relationships, and this process of unpacking can then be continued recursively, given that the contents of unpacked nodes themselves are placed in short-term memory as well, and hence are themselves subject to more detailed scrutiny, if so desired. (I suppose one could extend the fishing analogy by imagining that smaller fish are found in the stomach of the first fish caught, as it is "cleaned"—and so forth, recursively. But that fanciful and somewhat gory image is not crucial to my story.)

Thus, if it is placed under scrutiny, inside the "Frederikssund pier" node can be found nodes for the exchange of letters that preceded Robert's and my Danish reunion, for Frederikssund itself, for my Stockholm drive, for Robert's train trip, for a few of the town's streets and shops, for the pier, for my growing disappointment, and so on. Not all of these will be placed into short-term memory each time the event as a whole is recalled, nor will the inner structure of those nodes that *are* placed there necessarily be looked into, although it is quite possible that *some* of their inner structure will be examined.

Thus the unpacking process of this kind of high-level unlabeled node (such as the "Frederikssund-pier" node or the "Kolodner anagram" node) can fill short-term memory with a large number of interrelated structures. It must be stressed, however, that the unpacking process is highly context-dependent (i.e., sensitive to what concepts have been recently activated), and hence will yield a somewhat different filling-up of short-term memory on each occasion that the same high-level node is pulled up out of the ocean of long-term memory.

Once there are structures in short-term memory, then the perceptual process can be directed at any of them (this is, in fact, the kind of high-level perception that forms the core of the Copycat and Tabletop models of analogy-making—see [Hofstadter & FARG 1995]), the upshot of which will be the activation—thanks to analogy—of further nodes in long-term memory, which in turn causes new "fish" to be pulled out of that brine and placed into short-term memory's bucket. What we have described is, in short, the following *central cognitive loop:*

> A long-term memory node is accessed, transferred to short-term memory and there unpacked to some degree, which yields new structures to be perceived, and the high-level perceptual act activates yet further nodes, which are then in turn accessed, transferred, unpacked, etc., etc.

An Illustration of the Central Cognitive Loop in Action

THE FOREGOING may seem too abstract and vague, and so to make the ideas more concrete, I now will present a dialogue most of which actually took place, but some of which has been added on, so as to make some points emerge a little more clearly. The fact, however, that it all sounds perfectly normal is what matters—it certainly could pass for spontaneous cognition in the minds of two speakers. The dialogue exemplifies all the processes so far described, and—at least to my mind—shows how these processes are what drives thought. So here is the dialogue.

A and B are walking by a church when A looks up and notices that on the steeple, there are some objects that look like emergency-warning sirens attached to the base of the cross.

A: Hey, fancy that! Shades of "Praise the Lord and pass the ammunition!"

B: What do you mean?

A: Well, it's kind of amusing to me. On the one hand, the cross implies a belief in protection by the Lord, but on the other hand, the sirens suggest the need for a backup system, some kind of safety net. I mean, it's fine to believe in divine protection and all, but when it really comes to the crunch, religious people's true colors emerge . . .

B: Well, sooner safe than sorry, no?

A: Sure, but isn't a cross wrapped in danger sirens kind of hypocritical? I mean, why don't religious people put their money where their mouth is? If they really believe in God's benevolence, if they really have the courage of their own convictions, then how come it doesn't suffice to speak softly—why do they need to carry a big stick as well? Put it this way: Either you're a believer, or you ain't.

B: That's a bit black-and-white, isn't it?

A: Of course! As it should be! You can't have it both ways. Somehow this reminds me of when I had to leave my bags in a hotel in Italy for a few days, and the hotel people stored them in a tiny little chapel that was part of the hotel. A friend joked, "Well, this way they'll be protected." But why is such a remark so clearly a joke, even to religious people? Aren't churches houses of God? Shouldn't a *sacred* place be a *safer* place?

B: Yes, but being sacred doesn't make churches immune to disaster. We've all heard so often of churches whose roofs collapse on the assembled parishioners . . .

A: Exactly. And then pious people always say, "The Lord works in mysterious ways . . . It's beyond our comprehension." Well, how they can continue to believe after such an event is beyond *my* comprehension, that's for sure.

B: You're talking about people who claim to believe but in some sense act as if they don't really believe, deep down. But then there's the flip side of the coin: people who claim *not* to believe but act in a way as if they *do*. The reverse type of hypocrite, in short.

A: Do you have an example in mind?

B: Yes—Niels Bohr, the great Danish physicist. I once read that in his house there was a horseshoe hanging over one door, and someone asked him, "What's this all about?" Bohr answered, "Well, horseshoes are supposed to bring good luck, so we put it up there." The friend then said, "Come now—surely you don't *believe* it brings good luck, do you?" Bohr laughed and said, "Of course not!" And then he added, "But they say it works even if you *don't* believe in it."

A: I see your point—in a way Bohr's remark is the flip side of "Praise the Lord and pass the ammunition." In the trench-warfare case, you have a believer whose actions reveal deep doubts about their proclaimed belief, and in the Bohr case, you have a skeptic whose actions reveal that he may doubt his own skepticism. But that cross with the sirens—I just can't believe that they would wrap them around the *cross,* of all things—that's the height of irony! I mean, it's like some priest who's going into a dangerous area of town and doesn't just carry a handgun along in case of need, but in fact a *cross* that doubles as a handgun.

B: You've made the irony rather clear, I agree. But tell me—would you propose that the Pope, simply because he's a big-time believer in God, should travel through the world's cities without any protection? Would you propose that true believers, if they are to be self-consistent, shouldn't put locks on their churches?

A: Well, won't God take care of his flock? Especially the Pope?

B: It's not that simple.

A: Come on—if God doesn't look after the Pope, who *does* he look after?

B: Come on, yourself! They crucified Jesus, didn't they? If anyone should have had divine immunity, it was Jesus—but he didn't. And yet that in itself doesn't mean that Jesus wasn't God's son.

This exchange illustrates all of the themes so far presented. In the first place, it shows A and B using ordinary words—bite-size lexical items such as "cross," "sirens," "bags," "hotel," "when," "people," and dozens more—nouns, verbs, adjectives, adverbs, prepositions, and so forth. Nothing unusual here, of course, except that readers are being exhorted to picture each of these words as the tip of an iceberg that hides myriad hidden analogies—namely, the analogies that collectively allowed the category to come into being in the first place in the speaker's or listener's or reader's mind.

In the second place, the dialogue shows a good number of the shorter phrases cited in lists above being used in realistic situations—smallish stock phrases such as "Fancy that!," "kind of," "I mean," "ain't," "that's for sure," and many more. These phrases are used by speakers because they meet the rhetorical needs of the particular context, and when perceived by listeners they activate familiar rhetorical-context categories.

In the third place, the dialogue illustrates high-level perception—the retrieval of high-level labels for perceptions—such as A's opening statement, in which the lexical item "Praise the Lord and pass the ammunition" is the effortlessly evoked label for a church cross with warning sirens attached to it. In fact, all through the dialogue, the participants use large lexical items to label situations that are being categorized in real time in their minds. Thus we hear "backup system," "safety net," "when it really comes to the crunch," "flip side of the coin," "put your money where your mouth is," "sooner safe than sorry," "speak softly and carry a big stick," "black-and-white," and many more.

In the fourth place, we have large-scale remindings. First there is the shift from the cross-wrapped-in-sirens scene to the suitcases-left-in-hotel-chapel situation, then the shift to the collapsing-churches scenario, after which comes the shift, mediated by a kind of conceptual reversal, to Niels Bohr's horseshoe-that-works-despite-skepticism (probably an apocryphal story, by the way). Following that image comes a different kind of shift—an analogy where a given, known scenario is compared with a spontaneously concocted hypothetical scenario—thus, for instance, the cross-wrapped-in-sirens scene is compared with a hypothetical cross/handgun blend. This is swiftly followed by a trio of further concocted analogues: first the Pope traveling without protection, then churches that are left unlocked, and finally God not even taking care of his own son.

The Central Cognitive Loop in Isolation and in Interaction

THE BROAD-STROKE PATHWAY meandering through the limitless space of potential ideas during the hypothetical conversation of A and B is due to various actual scenes or imagined scenarios being reperceived, in light of recently activated concepts, in novel fashions and thereby triggering dormant memories, which are then fished up from dormancy to center stage (i.e., short-term memory), where, partially unpacked, they are in turn subjected to the exact same context-dependent reperception process. Around and around in such a loop, alternating between fishing in long-term memory and unpacking and reperceiving in short-term memory, rolls the process of cognition.

Note that what I have just described is not problem-solving, which has traditionally played such a large role in modeling of thought and been tightly linked with "analogical reasoning"; no, everyday thought is not problem-solving or anything that resembles it at all; rather, it is a nonrandom stroll through long-term memory, mediated by high-level perception (which is simply, to echo myself, another name for analogy-making).

To be sure, thought does not generally take place in a sealed-off vat or an isolation chamber; most of the time, external events are constantly impinging on us. Therefore the purely self-driven flow that the "central loop" would suggest is just half of the story—it is the contribution from within one's private cognitive system. The other half—the contribution from outside—comes from inanimate objects impinging on one's senses (skyscrapers and sunsets and splashes, for instance), from animate agents seen mostly as objects (mosquitos that one swats at, people that one tries not to bang into as one hastens down a crowded sidewalk), or from other cognitive agents (conversations with friends, articles read in the paper, e-mail messages, scenes in movies, and so on).

This buzzing, booming confusion in which one is immersed most of the time tends to obscure the constant running of the private inner loop—but when one retreats into solitude, when one starts to ponder or daydream, when one tries to close oneself off from these external impingements and to be internally driven, that is when the above-posited "central loop of cognition" assumes the dominant role.

Goal-Drivenness and the Central Loop

WHERE DO GOALS enter this picture? How does the deeply goal-driven nature of human thought emerge from what might seem to be the randomness of the posited central loop? The answer resides in the enormously biased nature of each individual's perception.

Each person, as life progresses, develops a set of high-level concepts that they tend to favor, and their perception is continually seeking to cast the world in terms of those concepts. The perceptual process is thus far from neutral or random, but rather it seeks, whenever possible, to employ high-level concepts that one is used to, that one believes in, that one is comfortable with, that are one's pet themes. If the current perception of a situation leads one into a state of cognitive dissonance, then one goes back and searches for a new way to perceive it. Thus the avoidance of mental discomfort—the avoidance of cognitive dissonance—constitutes a powerful internal force that helps to channel the central loop in what amounts to a strongly goal-driven manner.

The Whorf-Sapir Hypothesis: Language and the Central Loop

THE VIEWPOINT I have been proposing here—in most ways quite unrevolutionary!—can be rephrased in terms of "perceptual attractors," which are long-term mental loci that are zoomed into when situations are encountered (see [Kanerva 1988]). We all have many thousands of such attractors in our dormant memories, only a tiny fraction of which are accessed when we encounter a new situation. Where do such attractors come from? How public are they? Do they have explicit labels? Here I list three main types:

- standard lexical items (words, names, phrases, proverbs, etc.) provided to a vast public through a shared linguistic environment;
- shared vicarious experiences provided to a vast public through the media (e.g., places, personages, and events of small and large scale in books, movies, television shows, and so on), the smaller of which have explicit linguistic labels, the more complex of which have none;
- unique personal memories, lacking any fixed linguistic labels (such chunks are generally very large and complex, like the Frederikssund memory discussed above, or even far larger events, such as a favorite high-school class, a year spent in a special city, a protracted divorce, and so on)

Since a sizable fraction of one's personal repertoire of perceptual chunks is provided from without, by one's language and culture, this means that inevitably language and culture exert powerful, even irresistible, channeling influences on how one frames events. (This position is related to the "meme's-eye view" of the nature of thought, as put forth in numerous venues, most recently in [Blackmore 1999].)

Consider, for instance, such words as "backlog," "burnout," "micromanaging," and "underachiever," all of which are commonplace in today's America. I chose these particular words because I suspect that what they designate can be found not only here and now, but as well in distant cultures and epochs, quite in contrast to such culturally and temporally bound terms as "soap opera," "mini-series," "couch potato," "news anchor," "hit-and-run driver," and so forth, which owe their existence to recent technological developments. So consider the first set of words. We Americans living at the millennium's cusp perceive backlogs of all sorts permeating our lives—but we do so because the word is there, warmly inviting us to see them. But back in, say, Johann Sebastian Bach's day, were there backlogs—or more precisely, were backlogs perceived? For that matter, did Bach ever experience burnout? Well, most likely he did—but did he *know* that he did? Or did some of his Latin pupils strike him as being underachievers? Could he see this quality without being given the label? Or, moving further afield, do Australian aborigines resent it when their relatives micromanage their lives? Of course, I could have chosen hundreds of other terms that have arisen only recently in our century, yet which designate aspects of life that were always around to be perceived but which, for one reason or another, aroused little interest, and hence were neglected or overlooked.

My point is simple: we are prepared to see, and we see easily, things for which our language and culture hand us ready-made labels. When those labels are lacking, even though the phenomena may be all around us, we may quite easily fail to see them at all. The perceptual attractors that we each possess (some coming from without, some coming from within, some on the scale of mere words, some on a much grander scale) are the filters through which we scan and sort reality, and thereby they determine what we perceive on high and low levels.

Although this sounds like an obvious tautology, that part of it that concerns words is in fact a nontrivial proposition, which, under the controversial banner of "Sapir-Whorf hypothesis," has been heatedly debated, and to a large extent rejected, over the course of the twentieth century. I myself was once most disdainful of this hypothesis, but over time came to realize how deeply human thought—even my own!—is channeled by habit and thus, in the last accounting, by the repertoire of mental chunks (i.e., perceptual attractors) that are available to the thinker. I now think that it is high time for the Sapir-Whorf hypothesis to be reinstated, at least in its milder forms.

Language, Brains, and "Just Adding Water"

THE USUAL GOAL of communication is, of course, to set up "the same thought" in the receiver's brain as is currently taking place in the sender's brain. The mode by which such replication is attempted is essentially a drastic compression of the complex symbolic dance occurring in the sender's brain into a temporal chain of sounds or a string of visual signs, which are then absorbed by the receiver's brain, where, by something like the reverse of said compression—a process that I will here term "just adding water"—a new symbolic dance is launched in the second brain. The human brain at one end drains the water out to produce "powdered food for thought," and the one at the other end adds the water back, to produce full-fledged food for thought.

Take, for instance, the paragraph given a few pages back:

that time I spent an hour or two hoping that my old friend Robert, whom I hadn't seen in two years but who was supposed to arrive from Germany by train sometime during that summer day in the little Danish fishing village of

Frederikssund (which in a series of letters he and I had mutually picked out on maps, and in which I had just arrived early that morning after driving all night from Stockholm), might spot me as I lurked way out at the furthest tip of the very long pier, rather than merely bumping into me at random as we both walked around exploring the stores and streets and parks of this unknown hamlet

Obviously, this set of black marks on a white background is not similar to the time I spent in Frederikssund, nor is any part of it similar to a pier, a drive from Stockholm, a body of water, or dashed hopes. And yet these marks triggered in your brain a symbolic dance so vivid that you saw, in your mind's eye, a fishing village, two young friends, their joyful anticipation of a semi-random reunion, a pier stretching far out into a gulf, a barely visible person anxiously pacing at its tip, and so on. A never-before-danced dance inside your brain, launched by a unique set of squiggly shapes, makes you feel almost as if you had been there; had I spelled it out with another page or two of intricate black-on-white patterns, it would feel all the more vivid. This is a wonderful kind of transportation of ideas between totally different media—uprooting ideas from one garden and replanting them in a garden never even imagined before, where they flourish beautifully.

Transportation

IN HIS BOOK *The Poetics of Translation* [Barnstone 1993], poet and translator Willis Barnstone has a section called "The Parable of the Greek Moving Van," where he points out that on the side of all Greek moving vans is written the word μεταφορα (phonetically *metafora* and semantically "transportation"). He then observes:

> To come to Greece and find that even the moving vans run around under the sun and smog of greater Athens with advertisements for transportation, for metaphor, and ultimately with signs for TRANSLATION should convince us that every motor truck hauling goods from one place to another, every perceived *metamorphosis* of a word or phrase within or between languages, every decipherment and interpretation of a text, every role by each actor in the cast, every adaptation of a script by a director of opera, film, theater, ballet, pantomime, indeed every perception of movement and change, in the street or on our tongues, on the page or in our ears, leads us directly to the art and activity of translation.

I pack my mental goods down into tight, neat bundles, I load them as carefully as I can into the *metafora* truck of language, it drives from my brain to yours, and then you unpack. What a metaphor for communication! And yet it has often been said that all communication, all language, is metaphorical. Since I believe that metaphor and analogy are the same phenomenon, it would follow that I believe that all communication is via analogy. Indeed, I would describe communication this way: taking an intricate dance that can be danced in one and only one medium, and then, despite the intimacy of the marriage of that dance to that medium, making a radically new dance that is intimately married to a radically different medium, and in just the same way as the first dance was to its medium.

Trans-sportation

TO MAKE THIS ALL a little more concrete, let us consider taking a complex dance done in the medium of *the sport of basketball* and trans-sporting that dance into the rather different medium of *the sport of soccer*. Indeed, imagine taking the most enthralling basketball game you ever watched—perhaps a championship game you saw on television—and giving a videotape of that game to a "soccer choreographer," who will now stage all the details of an artificial soccer game that is in some sense analogous to your basketball game. Of course this could be done in many ways, some conservative and some daring.

Some choreographers, citing irreconcilable differences between the two sports (for instance, the difference in the number of players per team, the lack of any counterpart to a goalie in basketball, the low frequency of scoring in soccer relative to basketball, and on and on), might severely bend the rules of soccer, creating a game with only five players on a team, taking away the goalies, vastly reducing the size of the field (and the goals), and so forth, thus effectively creating a hybrid soccer-basketball game that looks very much like basketball, only it is played on grass and involves propelling the ball with the lower rather than the upper limbs. When one watched the reenactment of one's favorite basketball game in this artificial medium, one would not have the sense of watching a soccer game but of watching a very distorted basketball game.

Other choreographers, more willing to go out on a limb, would retain the normal rules of soccer but would attempt to stage a game whose every play *felt* like a particular play of the original basketball game, even though eleven players were retained on a side, even though the goals remained huge compared to baskets, even though there were still goalies, even though the goals might be coming a little too thick and fast, and so forth. There would be plays that would be essentially *like* slam-dunks while at the same time looking every bit like normal soccer plays. In such a case, one would feel one was watching a genuine soccer game—perhaps a peculiar one in some ways, but nonetheless genuine. In the ideal case, one could have the two counterpart games running on side-by-side television screens, and a "neutral" commentator using only terms that apply to both sports could be effectively heard as describing either of the games.

Anything in between these two extreme philosophies of "trans-sportation" can also be imagined—and just such a bizarre scenario is what I think everyday communication is actually like. Two brains are, in general, far more unalike than are the sports of soccer and basketball—and yet our society is predicated on mutual comprehensibility mediated by language.

Winding Up: On Associationism and the Cartesian Theater

THE CRUX OF this essay is the claim that thinking (at least when isolated from external influences) is a series of leaps involving high-level perception, activation of concepts in long-term memory, transfer to short-term memory, partial and context-dependent unpacking of chunks, and then further high-level perception, and so forth.

This may sound like no more than the age-old idea of associationism—that we think by jumping associatively from one thing to another. If that's all it came down to, my thesis would certainly be a sterile and vapid noncontribution to cognitive science. But the mechanisms I posit are more specific, and in particular they depend on the the transfer of tightly packed mental chunks from the dormant area of long-term memory into the active area of short-term memory, and on their being unpacked on arrival, and then scrutinized. Both transfer and perception are crucial, and in that respect, my thesis departs significantly from associationism.

Some readers, such as the author of *Consciousness Explained* [Dennett 1991], might feel they detect in this theory of thinking an insidious residue of the so-called "Cartesian theater"—a hypothetical theater in which an "inner eye" watches as various images go parading by on a "mental screen," and becomes "aware" or "conscious" of such imagery. Such a notion of thinking leads very easily down the slippery slope of nested homunculi, and thus to an infinite regress concerning the site of consciousness.

I would gladly plead guilty to the accusation of positing a "screen" upon which are "projected" certain representations dredged up from long-term memory, and I would also plead guilty to the accusation of positing an "inner eye" that scans that screen and upon it posts further representational structures, which trigger a descent via analogy into the dormant depths of long-term memory. I would insist, however, that the label "perception," as applied to what the "inner eye" does, be sharply distinguished from visual or any other kind of sensory perception, since in general it involves no sensory modality in any normal sense of the term (recall the perception of "ethnic cleansing" in a newspaper story). The nature of such abstract or high-level perceptual processing has been sketched out in work done by my students and myself over the years (see [Hofstadter & FARG, 1995]), and I will not attempt to describe it here. Clearly, since it has been implemented as a computer program (at least to a first approximation), such a model does not succumb to snagging on the fatal hook of infinite regress.

To those who would scoff at the very notion of any "inner screen" involved in cognition, I would point to the large body of work of perceptual psychologist Anne Treisman [e.g., Treisman 1988], which in my view establishes beyond any doubt the existence of temporary perceptual structures created on the fly in working memory (she calls them "object files")—a stark contrast to the connectionist-style thesis that all cognition takes place in long-term memory, and that it consists merely of simultaneous *conceptual activations* (possibly with attached temporal phases, so as to handle the "binding problem") without any type of transfer to, or structure-building in, a distinct working area. Although this more distributed view of the essence of cognition might appeal to opponents of the Cartesian theater, it does not seem to me that it comes anywhere close to allowing the richness of thought that back-and-forth flow between long-term and short-term memory would allow.

I hope that my speculative portrayal of analogy as the lifeblood, so to speak, of human thinking, despite being highly ambitious and perhaps somewhat overreaching, strikes a resonant chord in those who study cognition. My most optimistic vision would be that the whole field of cognitive science sud denly woke up to the centrality of analogy, that all sides suddenly saw eye to eye on topics that had formerly divided them most bitterly, and naturally—indeed, it goes without saying—that they lived happily ever after. Whatever.

BIBLIOGRAPHY

Barnstone, Willis (1993). *The Poetics of Translation: History, Theory, Practice*. New Haven: Yale University Press.

Becker, Joseph (1975). "The Phrasal Lexicon," in R. Schank and B. Nash-Webber, eds., *Theoretical Issues in Natural Language Processing*. Cambridge, MA: Bolt, Beranek and Newman.

Blackmore, Susan (1999). *The Meme Machine*. New York: Oxford University Press.

Dawkins, Richard (1976). *The Selfish Gene*. New York: Oxford University Press.

Dennett, Daniel C. (1991). *Consciousness Explained*. Boston: Little, Brown, and Company.

Hofstadter, Douglas R. (1997). *Le Ton beau de Marot*. New York: Basic Books.

Hofstadter, Douglas R. and David J. Moser (1989). "To Err Is Human; To Study Error-making Is Cognitive Science." *Michigan Quarterly Review*, vol. 28, no. 2, pp. 185–215.

Hofstadter, Douglas R. and the Fluid Analogies Research Group (1995). *Fluid Concepts and Creative Analogies*. New York: Basic Books.

Kanerva, Pentti (1988). *Sparse Distributed Memory*. Cambridge, Massachusetts: MIT Press (Bradford Books).

Schank, Roger (1982). *Dynamic Memory*. New York: Cambridge University Press.

Treisman, Anne (1988). "Features and Objects: The Fourteenth Bartlett Memorial Lecture." *Quarterly Journal of Experimental Psychology*, vol. 40A, pp. 201–237.

Revolutionary New Insoles Combine Five Forms of Pseudoscience

FROM *THE ONION*

If it sounds like science, it must be science—or so it sometimes seems in our tech-crazy, jargon-loving commercial culture. But as the pseudoscience writers at the pseudonewspaper The Onion *("America's Finest News Source") have figured out, the line between real and ersatz can get awfully blurry.*

MASSILLON, OH—Stressed and sore-footed Americans everywhere are clamoring for the exciting new MagnaSoles shoe inserts, which stimulate and soothe the wearer's feet using no fewer than five forms of pseudoscience.

"What makes MagnaSoles different from other insoles is the way it harnesses the power of magnetism to properly align the biomagnetic field around your foot," said Dr. Arthur Bluni, the pseudoscientist who developed the product for Massillon-based Integrated Products. "Its patented Magna-Grid design, which features more than 200 isometrically aligned Contour Points™, actually soothes while it heals, restoring the foot's natural bio-flow."

"MagnaSoles is not just a shoe insert," Bluni continued, "it's a total foot-rejuvenation system."

According to scientific-sounding literature trumpeting the new insoles, the

Contour Points™ also take advantage of the semi-plausible medical technique known as reflexology. Practiced in the Occident for over eleven years, reflexology, the literature explains, establishes a correspondence between every point on the human foot and another part of the body, enabling your soles to heal your entire body as you walk.

But while other insoles have used magnets and reflexology as keys to their appearance of usefulness, MagnaSoles go several steps further. According to the product's Web site, "Only MagnaSoles utilize the healing power of crystals to restimulate dead foot cells with vibrational biofeedback . . . a process similar to that by which medicine makes people better."

In addition, MagnaSoles employ a brand-new, cutting-edge form of pseudoscience known as Terranometry, developed specially for Integrated Products by some of the nation's top pseudoscientists.

"The principles of Terranometry state that the Earth resonates on a very precise frequency, which it imparts to the surfaces it touches," said Dr. Wayne Frankel, the California State University biotrician who discovered Terranometry. "If the frequency of one's foot is out of alignment with the Earth, the entire body will suffer. Special resonator nodules implanted at key spots in MagnaSoles convert the wearer's own energy to match the Earth's natural vibrational rate of 32.805 kilofrankels. The resultant harmonic energy field rearranges the foot's naturally occurring atoms, converting the pain-nuclei into pleasing comfortrons."

Released less than a week ago, the $19.95 insoles are already proving popular among consumers, who are hailing them as a welcome alternative to expensive, effective forms of traditional medicine.

"I twisted my ankle something awful a few months ago, and the pain was so bad, I could barely walk a single step," said Helene Kuhn of Edison, NJ. "But after wearing MagnaSoles for seven weeks, I've noticed a significant decrease in pain and can now walk comfortably. Just try to prove that MagnaSoles didn't heal me!"

Equally impressed was chronic back-pain sufferer Geoff DeAngelis of Tacoma, WA.

"Why should I pay thousands of dollars to have my spine realigned with physical therapy when I can pay $20 for insoles clearly endorsed by an intelligent-looking man in a white lab coat?" DeAngelis asked. "MagnaSoles really seem like they're working."

Brilliant Light

FROM *THE NEW YORKER*

Years before he became a famous neurologist and the author of such classic books as Awakenings *and* The Man Who Mistook His Wife for a Hat, *Oliver Sacks dreamed of a different profession. During and immediately following the time he was evacuated from London during World War II (a period he calls his "exile"), the young Sacks was nurtured in the beauty and majesty of science by his family. His Auntie Len encouraged his love of math, and his Uncle "Tungsten," so called because he manufactured tungsten-filament light bulbs (but whose real name was Dave), introduced him to the mysterious properties of metals and other elements. In this memoir he affectionately recalls his days as a childhood experimenter who wanted more than anything to be a chemist when he grew up.*

For me, the refuge I found at first was in numbers. My father was a whiz at mental arithmetic, and I, too, even at the age of six, was quick with figures—and, more, in love with them. I liked numbers because they were solid, invariant; they stood unmoved in a chaotic world. There was in numbers and their relation something absolute, certain, not to be questioned, beyond doubt. (Years later, when I read *1984*, the climactic horror for me, the ultimate sign of Winston's disintegration and surrender, was his being forced, under torture, to deny that two and two is four. Even more terrible was the fact that eventually he began to doubt this in his own mind—that, finally, numbers failed him, too.)

I particularly loved prime numbers, the fact that they were indivisible, could not be broken down, were inalienably themselves. (I had no such confidence in myself, for I felt I was being divided, alienated, broken down more and more every week.) Why did primes come when they did? Was there any pattern, any logic to their distribution? Was there any limit to them, or did they go on forever? I spent innumerable hours factoring, searching for primes, memorizing them. They afforded me many hours of absorbed, solitary play, in which I needed no one else.

I made a grid, ten by ten, of the first hundred numbers, with the primes blacked in, but I could see no pattern, no logic to their distribution. I made larger tables, increased my grids to twenty squared, thirty squared, but still could discern no obvious pattern.

The only real holidays I had during the war were visits to a favorite aunt in Cheshire, in the midst of Delamere Forest, where she had founded the Jewish Fresh Air School for "delicate children." All the children, indeed, had little gardens of their own, squares of earth a couple of yards wide, bordered by stones. I wished desperately that I could go there, rather than Greystone—but this was a wish I never expressed (though I wondered if my clear-sighted and loving aunt did not divine it).

On my visits, Auntie Len always delighted me by showing me all sorts of botanical and mathematical pleasures. She showed me the spiral patterns on the faces of sunflowers in the garden, and suggested I count the florets in these. As I did so, she pointed out that they were arranged according to a series—1, 1, 2, 3, 5, 8, 13, 21, etc.—each number being the sum of the two that preceded it. And if one divided each number by the number that followed it (1/2, 2/3, 3/5, 5/8, etc.), one approached the number 0.618. This series, she said, was called a Fibonacci series, after an Italian mathematician who had lived centuries before. The ratio of 0.618, she added, was known as the Divine Proportion or Golden Section, an ideal geometrical proportion found in many plants and shells, and often used by architects.

She would take me for long, botanizing walks in the forest, where she had me look at fallen pinecones, to see that they, too, had spirals based on the Golden Section. She showed me horsetails growing near a stream, had me feel their stiff, jointed stems, and suggested that I measure these when I got back to school, and plot the lengths of the successive segments as a graph. When I did so, and saw that the curve flattened out, she explained that the increments were "exponential," and that this was the way growth usually occurred. These ratios, these geometric proportions, she told me, were to be found all over nature—numbers were the way the world is put together. Numbers, my aunt said, are the way God thinks.

The association of plants, of gardens, with numbers assumed a curiously intense, symbolic form for me. I started to think in terms of a kingdom or a realm of numbers, with its own geography, languages, and laws; but, even more, of a garden of numbers, a magical, secret, wonderful garden in which I could wander and play to my heart's content. It was a garden hidden from, inaccessible to, the bullies and the headmaster; and a garden, too, where I somehow felt welcomed and befriended. Among my friends in this garden were not only primes and Fibonacci sunflowers but perfect numbers (such as 6 or 28, the sum of their factors excluding themselves); Pythagorean numbers, whose square was the sum of two other squares (such as 3, 4, 5 or 5, 12, 13); and "amicable numbers"—pairs of numbers (such as 220 and 284) in which the factors of each added up to the other. And my aunt had shown me that my garden of numbers was doubly magical—not just delightful and friendly, always there, but part of the plan on which the whole universe was built.

Uncle Tungsten

I RETURNED to London in the summer of 1943, after four years of exile, a ten-year-old boy, withdrawn and disturbed in some ways but with a passion for metals, for plants, and for numbers.

I delighted in being able to visit Uncle Tungsten again, and I think he also delighted in having his young protégé back, for he would spend hours with me in his factory and his lab, answering questions as fast as I could ask them.

He had several glass-fronted cabinets in his office, one of which contained a series of electric light bulbs: there were several Edison bulbs from the early eighteen-eighties, with filaments of carbonized thread; a bulb from 1897, with a filament of osmium; and several bulbs from a few years later, with spidery filaments of tantalum tracing a zigzag course inside them.

Then there were the more recent bulbs—these were Uncle Dave's special pride and interest, for some he had pioneered himself—with tungsten filaments of all shapes and sizes. There was even one labelled "Bulb of the Future?" It had no filament, but the word "Rhenium" was inscribed on a card beside it.

I had heard of platinum, but the other metals—osmium, tantalum, rhenium—were new to me. Uncle kept samples of them all, and some of their ores, in a cabinet next to the bulbs. As he handled them, he would expatiate on their unique, sovereign properties and qualities, how they had been discovered, how they were refined, and why they were so suitable for making filaments.

He would bring out a pitted gray nugget: "Dense, eh?" he would say, tossing it to me. "That's a platinum nugget. This is how it is found, as nuggets of pure metal. Most metals are found as compounds with other things, in ores. There are very few other metals which occur native like platinum—just gold, silver, and copper, and one or two others." These other native metals had been known, he said, for thousands of years, but platinum had been "discovered" only two hundred years ago, for though it had been prized by the Incas for centuries, it was unknown to the rest of the world. When the explorers brought it back, in the eighteenth century, the new metal enchanted all of Europe—it was denser, more ponderous than gold, and, like gold, it was "noble" and never tarnished. It had a lustre exceeding that of silver. (Its Spanish name, *platina*, meant "little silver.")

Native platinum was often found with two other metals, iridium and osmium, which were even denser, harder, more refractory. Here Uncle pulled out samples for me to handle, mere flakes, no larger than lentils, but astoundingly heavy. These were "osmiridium," a natural alloy of osmium and iridium, the two densest substances in the world. There was something about heaviness, density—I could not say why—which gave me a thrill, and an immense sense of security and comfort. Osmium, moreover, had the highest melting point of all the platinum metals, so it was used at one time, Uncle said, to replace the platinum filaments in light bulbs.

The great virtue of the platinum metals was that they were as noble and workable as gold but had much higher melting points. Crucibles made of platinum could withstand the hottest temperatures; beakers and spatulas of it could withstand the most corrosive acids. Uncle Dave often used platinumware in his own lab, sometimes alloyed with other platinum metals to give it greater hardness, and a still higher melting point. He pulled out a small crucible from the cabinet, beautifully smooth and shiny. It looked new. "This was made around 1840," he said. "A century of use, and almost no wear."

UNCLE DAVE SAW the whole earth, I think, as a gigantic natural laboratory, where heat and pressure caused not only vast geologic movements but innumerable chemical miracles. "Look at these diamonds," he would say. "They were formed thousands of millions of years ago, deep in the earth, under unimaginable pressures."

He liked to pull out the native metals from his cabinet—twists and spangles of rosy copper, wiry silver, latticed gold. "Think how it must have been," he said, "seeing metal for the first time—sudden glints of reflected sunlight, sudden shinings in a rock or at the bottom of a stream!"

He would conjure up the first smelting of metal, how cavemen might have used rocks containing a copper mineral—green malachite, perhaps—to surround a cooking fire and suddenly realized as the wood turned to charcoal that the green rock was bleeding, turning into a red liquid: molten copper.

It took a much hotter fire, a white-hot fire, to obtain tungsten. Uncle Dave handed me a little ingot. "Tungsten," he said. "No one realized at first how perfect a metal tungsten was. It has the highest melting point of any metal, it is tougher than steel, and it keeps its strength at high temperatures—an ideal metal!"

Uncle had a variety of tungsten bars in his office—some he used as paperweights, but others had no discernible function whatever, except to give pleasure to their owner and maker. And, indeed, by comparison steel bars, and even lead, felt light and somehow porous, tenuous. "These lumps of tungsten have an extraordinary concentration of mass," he would say. "They would be deadly as weapons—far deadlier than lead."

But sooner or later Uncle's soliloquies and demonstrations before the cabinet all returned to tungsten's mineral ores. One of these, scheelite, was named after the great Swedish chemist Carl Wilhelm Scheele, who was the first to show that it contained a new element. The ore was so dense that miners called it "heavy stone," or *"tung-sten,"* the name subsequently given to the element itself. Scheelite was found in beautiful orange crystals that fluoresced bright blue in ultraviolet light. Uncle Dave kept specimens of scheelite and other fluorescent minerals in a special cabinet in his office. The dim light of Farrington Road on a November evening, it seemed to me, would be transformed when he turned on his Wood's lamp, and the luminous chunks in the cabinet suddenly glowed orange, turquoise, crimson, green.

Though scheelite was the largest source of tungsten metal, the metal had first been obtained from a different mineral, called wolframite. Indeed, tungsten was sometimes called wolfram, and still retained the chemical symbol of W. This thrilled me, because my own middle name was Wolf. Heavy seams of the tungsten ores were often found along with tin ore, and the tungsten made it more difficult to isolate the tin. This was why, my uncle continued, they had originally called the metal wolfram—for, like a hungry animal, it "stole" the tin. I like the name wolfram, its sharp, animal quality, its evocation of a ravening, mystical wolf—and thought of it as a tie between Uncle Tungsten, Uncle Wolfram, and myself, O. Wolf Sacks.

NAMES FASCINATED ME—their sounds, their associations, the sense they gave of people and places. The names of the elements were filled with such evocations, but there were only a few dozen of these, whereas the number of minerals ran into the hundreds or thousands. These were all beautifully laid out, with their names and formulas, in the cabinets of the Geology Museum, in South Kensington, where, later, I would go whenever I could.

The older names gave one a sense of antiquity and alchemy: corundum and galena, orpiment and realgar. (Orpiment and realgar, two arsenic sulfides, went euphoniously together, and made me think of an operatic couple, like Tristan and Isolde.) There was pyrites, fool's gold, in brassy, metallic cubes, and chalcedony and ruby and sapphire and spinel. Zircon sounded Oriental; calomel, Greek—its honeylike sweetness, its "mel," belied by its poisonness. There was the medieval-sounding sal ammoniac. There was cinnabar, the heavy red sulfide of mercury, and massicot and minium, the twin oxides of lead.

Then there were minerals named after people. One of the most common minerals, much of the redness of the world, was the hydrated iron oxide called goethite. (Was this named in honor of Goethe, or did he discover it? I had read that he had a passion for mineralogy and chemistry.) Many minerals were named after chemists—gaylussite, scheelite, berzelianite, bunsenite, liebigite, moissanite, crookesite, and the beautiful, prismatic "ruby-silver," proustite. There was samarskite, named after a mining engineer, Colonel Samarski. There were other names that were evocative in a more topical way: stolzite, a lead tungstate, and scholzite, too. Who were Stolz and Scholz? Their names seemed very Prussian to me, and this, just after the war, evoked an anti-German feeling. I imagined Stolz and Scholz as Nazi officers with barking voices, swordsticks, and monocles.

Other names appealed to me mostly for their sound, and for the images they conjured up. I loved classical words and their depiction of simple properties—the crystal forms, colors, shapes, and optics of minerals—like diaspore and anastase and microlite and polycrase. A great favorite was cryolite—ice stone, from Greenland, so low in refractive index that it was transparent, almost ghostly, and like ice, became invisible in water.

ON ONE VISIT, Uncle Dave showed me a large bar of aluminum. After the dense platinum metals, I was amazed at how light it was, scarcely heavier than a piece of wood. "I'll show you something interesting," he said. He took a smaller lump of aluminum, with a smooth, shiny surface, and smeared it with mercury. All of a sudden—it was like some terrible disease—the surface broke down, and a white substance like a fungus rapidly grew out of it, until it was a quarter of an inch high, then half an inch high, and it kept growing and growing until the small piece of aluminum was completely eaten up. "You've seen iron rust, oxidizing, combining with the oxygen in the air," Uncle said. "But here, with the aluminum, it's a million times faster. That big bar is still quite shiny, because it's covered by a fine layer of oxide, and that protects it from further change. But rubbing it with mercury destroys the surface layer, so then the aluminium has no protection, and it combines with the oxygen in seconds."

I found this magical, astounding, but also a little frightening—to see a bright and shiny metal reduced so quickly to a crumbling mass of oxide. It made me think of a curse or a spell, the sort of disintegration I sometimes saw in my dreams. It made me think of mercury as evil, as a destroyer of metals. Would it do this to every sort of metal?

"Don't worry," Uncle answered. "The metals we use here, they're perfectly safe. If I put this little bar of tungsten in the mercury, it would not be affected at all. If I put it away for a million years, it would be just as bright and shiny as it is now." In a precarious world, tungsten, at least, was stable.

AS THE YOUNGEST of almost the youngest (I was the last of four, and my mother the sixteenth of eighteen), I was born nearly a hundred years after my maternal grandfather, and never knew him. He was born Mordechai Fredkin, in 1837, in a small village in Russia. As a youth, he managed to avoid being impressed into the Cossack Army, and fled Russia, using the passport of a dead man named Landau. He was sixteen. As Marcus Landau, he made his way to Paris, and then Frankfurt, where he married. (His wife was sixteen, too.) Two years later, in 1855, now with the first of their children, they moved to England.

My mother's father was, by all accounts, a man drawn equally to the spiritual and the physical. He was by profession a boot and shoe manufacturer, a *shochet* (a kosher slaughterer), and later a grocer—but he was also a Hebrew scholar, a mystic, an amateur mathematician, and an inventor. He had a wide-ranging mind: he published a newspaper, the *Jewish Standard*, in his basement, from 1888 to 1891; he was interested in the new science of aeronautics, and corresponded with the Wright brothers, who paid him a visit when they came to London in the early nineteen-hundreds. (Some of my uncles could still remember this.) He had a passion, my aunts and uncles told me, for intricate arithmetical calculations, which he would do in his head, while lying in the bath. But he was drawn, above all, to the invention of lamps—safety lamps for mines, carriage lamps, street lamps—and he patented many of these in the eighteen-seventies. When I was very small, my mother would take me to the Science Museum, in South Kensington, up to the top floor, where there was a simulacrum of a coal mine, its dim, low passages lit by fitful beams. There she would show me the Landau safety lamp on display, right next to the more famous Humphry Davy lamp.

A polymath and an autodidact himself, Grandfather was passionately keen on education—and, most especially, a scientific education—for all his children, for his nine daughters no less than for his nine sons. Whether it was this or the sharing of his own passionate enthusiasms, seven of his sons were eventually drawn to mathematics and the physical sciences—including the two I was closest to, Uncle Dave and Uncle Abe.

Stinks and Bangs

MY PARENTS and my brothers had introduced me, even before the war, to some kitchen chemistry: pouring vinegar on a piece of chalk in a tumbler and watching it fizz; then pouring the heavy gas this produced, like an invisible cataract, over a candle flame, putting it out straightaway. Or taking red cabbage, pickled with vinegar, and adding household ammonia to neutralize it. This would lead to an amazing transformation, the juice going through all sorts of colors, from red to various shades of purple, to turquoise and blue, and finally to green. I enjoyed these experiments, I wondered what was going on, but I did not feel a real chemical passion—a desire to compound, to isolate, to decompose, to see substances changing, familiar ones disappearing and new ones appearing in their stead—until I returned from Greystone, remet Uncle Dave, and saw his lab and his passion for experiments of all kinds.

Now, after hearing him talk about chemistry, and starting to read about chemistry and chemists myself, I longed to have a lab of my own.

As a start, I wanted to lay hands on cobaltite and niccolite, and compounds or minerals of manganese and molybdenum, of uranium and chromium—all those wonderful elements that were discovered in the eighteenth century. I wanted to pulverize them, treat them with acid, roast them, reduce them—whatever was necessary—so I could extract their metals myself. I knew, from looking through a chemical catalogue at the factory, that one could buy these metals already purified, but it would be far more fun, far more exciting, I reckoned, if I was able to make them myself. This way, I would enter chemistry, start to discover it for myself, in much the same way its first practitioners did—I would live the history of chemistry in myself.

It was through reading Mary Elvira Weeks's "Discovery of the Elements"— a book published just before the war—that I got a vivid idea of the lives of many chemists, the great variety, and sometimes vagaries, of character they showed; and the relation (sometimes) between their characters and their work. Here I found quotations from the early chemists' letters, which portrayed their excitements (and despairs) as they fumbled and groped their way to their discoveries, losing the track now and again, getting caught in blind alleys, though ultimately reaching the goal they sought.

If Humphry Davy was the first chemist I had ever heard of, he was also the one I most warmed to. I loved reading of his experiments with explosives and electric fish; his discovery of incandescent metal filaments and electric arcs; of catalysts (it was only now that I realized why we had a platinum loop above the gas stove); of the physiological effects of nitrous oxide, laughing gas; and, above all, of his using the just invented electric battery to isolate the alkali and alkaline-earth metals in a single miraculous year. He appealed to me especially because he was boyish and impulsive, the way he danced with joy all around his lab when he first isolated potassium, in 1807, and saw the shining metallic globules burst and take fire. Davy moved me to emulation—to sampling the effects of nitrous oxide for myself (my mother kept a cylinder of it in her obstetric bag), and making my own sodium and potassium by electrolysis.

I was awed, too, by the figure of Mendeleev—his passionate search for order among the elements (more than sixty were known by the eighteen-fifties, a rich chaos), and his final discovery of such an order (supposedly in a dream) in 1869. When I first saw the Periodic Table, it hit me with the force of revelation—it embodie;d, I was convinced, eternal truths, the eternal and necessary order of the elements. I thought of Mendeleev as a sort of Moses, bearing the tablets of the God-given Periodic Law.

And then, in a different mode, there was Marie Curie, who had spent four backbreaking years in a shed extracting a pinch of "her" element, radium, from four stubborn tons of pitchblende. Radium, my mother would say, was a magical element, with unique powers to harm and cure. She herself had worked with radium therapy at the Marie Curie Hospital, in London, and had met Marie Curie on one occasion. (I was intrigued when she told me of the radium "bomb" at the hospital, and the fine gold needles full of radon which were inserted into tumors.) Eve Curie's biography of her mother—which my mother gave me when I was ten—was the first portrait of a scientist I read, and Marie Curie was added to my pantheon of heroes. (Fifty-five years later, in

1998, at a meeting in New York to celebrate the centenary of the Curies' discovery of polonium and radium, I met Eve Curie, now in her nineties, and asked her to sign the book.)

It was through reading these accounts that I first realized one could have heroes in real life. There seemed to me an integrity, an essential goodness, about a life dedicated to science. I had never given much thought to what I might be when I was "grown up"—growing up was hardly imaginable—but now I knew: I wanted to be a chemist. A sort of eighteenth-century chemist coming fresh to the field, looking at the whole, undiscovered world of natural substances and minerals, analyzing them, plumbing their secrets, finding the wonder of new and unknown metals.

Lab Notes

FROM *HARPER'S* MAGAZINE

If chemistry was a magical refuge for the young Oliver Sacks, it was any-thing but for the young Don Asher. Like Sacks, Asher grew up in a family of science professionals, but rather than dreaming of silicon and mercury, his thoughts drifted to Ellington and Calloway. In this entertaining mem-oir, Asher, now a writer and jazz pianist, describes how he most emphati-cally fell out of love with science.

Had silicon been a gas I would have been a major general.
   —JAMES MCNEILL WHISTLER,
     looking back from the vantage of celebrity on his departure
     from West Point in 1854, after flunking chemistry

When I graduated from high school the Big Band era was in full swing. The pop tunes of the day were "Full Moon and Empty Arms," "How Are Things in Glocca Morra?" and the lovely "Stella by Starlight." The male members of my family—cousins and uncles in-cluded—had all gone (or were on their way) to college and prepared for suit-able professions, so I was feeling that heat. But I had no mission and heard no suitable clarion calls. What I dearly wanted to be was a nightclub piano player, which hardly required a college degree. In the Forties the scions of middle-class Jewish families did not become ivory pushers. Lacking socially acceptable goals, we followed in designated footsteps. My father had an artistic soul but earned his living as a chemist with the Worcester (Mass.) Water Works. A brilliant, troubled, melancholic man, he committed suicide when I was three. Through childhood, close watch was kept on me for the malign strain; my mother considered me the more emotionally vulnerable of her two sons and, by extension, the more likely to inherit my father's fragile psyche. Older brother Herbie, a sturdy straight-arrow, had followed my father to Worcester Polytech and accepted a chemical engineering position with a plastics company in Framingham.

Dutiful son and kid brother, I enrolled at Cornell with a major in organic chemistry. The school's main campus was truly impressive, situated on a broad promontory overlooking Cayuga Lake and carved by roaring gorges, its ivied stone halls shaded by majestic elms. The chemistry department, housed in massive Baker Lab, was considered world-class, and I set out in an optimistic, even visionary mode, Einsteinian stars in my eyes, contemplating future major addresses before eminent scientific bodies. *Gentlemen: I should like to propose a structure for the potassium salt of ribonucleic acid with a view toward explaining certain aspects of growing genetic interest . . .*

Baker Lab was a museum, an exploratorium of exotic effluvia; scents clung like damp, like incense, slipping around doorjambs, seeding corridors. Sweetish perfume of unsaturated ester; head-snapping ethers and amines; the vinegary (and glacial) acetic acid; essence of almonds, cinnamon, camphor; fragrances redolent of fresh linoleum varnish, the sun-warm pinewoods of childhood; other scents more stringent, insidious—decaying-fruit exhalation of ketone, sting of aldehyde, the fuming and corrosive acids. Amid the cool, slippery soapstone counters, the gleaming glassware and furiously bubbling flasks, one could believe the choice wasn't all that bad; there were worse ways to spend one's life if one lacked the lineage and finesse for the diplomatic corps, the cold nerve of a surgeon, the chutzpah of a lawyer. Small magic shows daily: flocculent lavender-white-rose precipitates materialized from the mixture of two spring-clear liquids. Release a glittering pearl drop into a murky solution and *sha-boom!* all clear, Tinker to Evers to Chance, the dear little molecules reacting on cue with one another, punctiliously, according to law; acetone, all-purpose voodoo elixir, reagent, solvent, cleaner of delicate glassware; thin-lipped beakers, reflux condensers—proud erect sentinels clamped to their iron stands; the graduated burettes and androgynous one-, two-, and three-necked flasks. . . . A luxurious chromatograph of colors: white-on-white silver chloride, cool green of copper salts, the sable-black of certain sulfides.

Pass a slender wire dipped in potassium salt solution through the spurting blue flame of a Bunsen burner and *whooomm*, violet-blue light flashed and crackled like lightning off a space-probe antenna. Formidable challenges loomed, exhilarating years ahead, probing the secrets of a staggering array of sinuous, branching, crookback molecules. *Gentlemen, I am pleased to inform you of my decision to accept the position of assistant research director at your Wilmington plant....*

Occasionally experiments just plain didn't work out. Pristine fluffy-white precipitates would turn out to be worthless salt; cheerfully simmering ponds of liquid would grow inexplicably surly and bilious within their flasks, and without warning, goaded to a sudden boil-like soreness, spew a stream of foaming liquid out the top of the condenser like a hot-spring geyser.

During preadolescence—I was a loopy kid, a nebbish, afraid of shadows (relatives attributed this to fallout from my father's early death, the shock of abandonment and absence of a paternal rudder, though I can't say I felt abandoned; I scarcely remembered him)—I habitually kicked the houses of neighborhood kids I was mad at and cravenly avoided fistfights, protecting the hands that were just beginning to explore the keyboard's mysteries. These impulses attenuated over the years but never completely vanished, and I'd now find myself kicking the foundation of the soapstone counter as if to jolt the balky molecules, visions of scientific eminence gone glimmering, so much dry-ice smoke. But just when I'd grow really disheartened, the future as murky as a mill-town sky, an experiment would work out with undeserved felicity, the vile and noxious sulfides suddenly smelling like night-blooming jasmine. Joy!

"Very neat, Mr. Asher." Gangly Dr. Rodman, my patient mentor, leaning over my shoulder.

"When you're hot you're hot, Professor," I murmured modestly.

I joined a fraternity. One of my brothers was a jazz-freak amateur drummer. We spent our weekend nights hanging out at the Green Lantern, a funky cellar dive by the railroad tracks in downtown Ithaca that featured a gutty trumpet-sax-piano-drums combo and a beanpole stripper, Cherry Picker (six feet two in spike heels and silver tiara), who strutted and peeled her way across the board floor to medium-tempo Fats Waller and an obbligato of pounding beer mugs and crazed rebel yells from polluted locals. The band let us sit in one at a time during dance sets, allowing me to expand the fledgling improvisational skills I'd picked up in Worcester's lakeside saloons on the sly during high school. I loved that fragrant joint, mingled aromas of spilt lager, sawdust, and cigarettes, the slow-drag dancers and statuesque stripper gliding through smoky gin-mill light. Seduced by the good vibes and down-and-dirty music, I began to drink a fair amount: draft brew, an occasional boilermaker, and an evil concoction of my own design—Southern Comfort and root beer.

Furthering my education in this regard was good-time Charlie Solinsky, a doctoral candidate in physical chemistry. Charlie had his own private lab on the top floor of Baker and presided over biweekly cocktail parties for favored grad and undergrad students. We chipped in for munchies and took turns double-distilling C_2H_5OH (ethyl alcohol) straight from the reagent shelf, then mixing it with cranberry or grapefruit juice. These libations, iced and sipped sweetly from 250-cc. beakers, could be hairy indeed, depending on who was in charge of the distillery.

It was on a morning following one of these crapulous sessions that I first began to question my career choice. In the small lab I shared with a classmate I'll call Eunice, an ingenious, built-from-scratch apparatus burst at the seams, liberally spattering Eunice's open lab coat and cotton dress with sulfuric acid. I saw her totter backward, walleyed, and freeze. I ran around the counter, pushed her into the open shower stall—a provision of most labs—and pulled the link chain. Through the tumultuous downpour I watched her smoking garments dissolve. Spluttering, face crumpled, she stood sopping wet but un-scathed in tattered dress, slip, and brassiere, arms crossed over her meager chest, beseeching me with her eyes to leave her a shred of privacy.

Two weeks later, I found myself falling to the concrete floor as if poleaxed—I can still hear the distant library-tower chimes peeling ten as I dropped—and lay there thrashing and gasping like a fish in a skiff. I knew in-stantly why I was down. I had mindlessly stuck my nose in a flask collecting distillate—a treacherous, highly volatile, low-molecular-weight amine—which closed off my oxygen supply as efficiently as a down pillow. No one else was around; Eunice had left a half hour earlier for the building's library. Although physically incapacitated, my oxygen-deprived brain still whirred. The culprit molecule was a base; what I desperately needed in order to survive was a quick whiff of a mild acid to neutralize it. The antidote and my passport to life, a glass-stoppered bottle of glacial acetic acid, sat on the reagent shelf above my head. Arms flailing, sucking what I felt were my final breaths, I could neither rise nor cry out. What unreeled before my eyes now was not a swift synopsis of my nineteen years of life but Cherry Picker shimmying across the boards, naked but for tiara and spike heels. A shadow loomed in the lab doorway. Oxygen-starved, I hallucinated the Angel of Death. My sights were too high: it was Eunice, resurfacing like a vision of the Virgin Mary.

Her mouth gaped as I flopped, blue-faced, on the floor. With my penultimate breath I pointed soundlessly to the reagent shelf. She grabbed a bottle of acetone. *No, no.* One hand clawing my throat, I pointed *left, one over.* She moved tentatively for the acetic acid—I nodded violently *yes.* Dear intelligent Eunice unstoppered the bottle and brought it to my nose. Extract of the gods, *elixir vita.* Sinuses and contiguous passages magically cleared. I grasped the counter and feebly pulled myself to my knees. Wheezing, I clasped diminutive, homely, lab-coated Eunice to me as if she were Veronica Lake in a polka-dot bikini.

IT WASN'T ALL FUME AND DOOM. As an upperclassman I lured many a house-party date to my top-floor room at Alpha Epsilon Pi with a promise of simulated atomic fission—a kind of pyrotechnic showbiz foreplay—making good my commitment by raising the window and dropping a punctured can packed with metallic sodium into the boiling waters of Cascadilla Gorge 150 feet below. Have you ever watched metallic sodium hit water? *Ka-boom.* Krakatau revisited, shades of Los Alamos in upstate New York. A hard act to follow. As much as a school could swing during the Truman years, Cornell swung. Nearly a hundred fraternities and sororities held an endless succession of weekend blasts, campus wetter than Cayuga's waters thanks to enlightened New York legislators. I have fervent memories of pounding many a parlor grand into submission in the wee hours (they were painted peach, pink, and powder blue in the tonier sororities), fueled by orange blossoms and daiquiris. Skoal! Halycon days, Cornell!

My brother Herbie had recently changed jobs and was now working at the famed Corning Glass Works, an hour's drive from Cornell. On a Saturday morning in May, near the close of my junior year, he came to check on me. He'd had a flat tire on the way and arrived at my room flushed and sweaty, a dark greasy stain on one knee of his wide tan slacks.

"I've got just the stuff to clean that," I told him. "Take 'em off."

"Hold your horses, let's try rubbing alcohol first. If that doesn't—"

"Herbie, I've got the right stuff, trust me. Gimme the pants and I'll be right back." I left him sitting plumply in his jockey shorts and hurried down the hall to the can, where I had stored in my corner of the communal cabinet a bottle of acetone appropriated from Baker Lab. I'd become enamored of this super solvent: dump 100 cc. in a gunk-caked flask, swirl it around, and *shazaam!*— clear as a swami's crystal ball. Using a rough-textured hand towel, I ministered to bro's pants, returned to the room, and hung them on the inside hook of the closet door.

"This stuff in quantity could've dissolved the La Brea tarpits. It'll dry in no time. Then I'll show you the campus."

"Are you eating okay? You're skinny as a nail."

"Always have been." I glanced at his thighs bursting from the jockey shorts.

"If you're tight for money just say the word."

"I'm straight. I've been making extra loot playing two nights a week at a town bar." I'd temporarily replaced the Green Lantern's pianist, who'd sprained a couple of fingers in an after-hours brawl with an excitable patron.

"I hope it's safer than some of the Worcester joints you—"

Herbie's attention had been drawn to his pants, visible in the open closet door. Now he abruptly pushed out of the chair and walked over to inspect them.

"For Christ sake, kid . . ."

"What?" I came up behind him.

"What'd you use, fuming hydrofluoric acid?" Where I had applied the solvent was a moldering threadbare patch the size of a drink coaster. The fabric was disintegrating before our eyes.

"Acetone," I said weakly.

Herbie squeezed his eyes shut and blew out his breath. "You dumb bunny, those slacks are a soluble acetate rayon!"

"Ah."

"Jesus, kid, where's your head? Couldn't you see what the fabric was? A clerk in a clothing store with rudimentary knowledge of synthetic materials knows you don't use—I mean this doesn't bode great for your future." He lifted the savaged pants tenderly between thumb and forefinger.

"You can use a pair of mine."

"How do you propose I squeeze into them—with scissors and a shoehorn?"

I furtively kicked the base of the wall. "Listen, I'll pay for the pants."

"What're you talking, they're cheapo—ten, twelve bucks. The point is—" He blew out his breath again, slowly. "Your chem department here is supposed to be topnotch. I mean, isn't anything rubbing off? Ah . . . never mind, let it go." He lightly patted my shoulder. "Come on, show me your campus. So I'll look like a slob, won't be the first time." He sat down and pulled on the disintegrating slacks, shaking his head. "But if I were you, kid"—his tongue had popped into his cheek; I could hear the soft laughter huffing in his throat—"I'd keep right on practicing the piano."

GRADUATION DAY (not a bad tune). Corporate interviewers by the dozens infiltrated the various campuses. It was a seller's market; I entertained unreasonably high hopes. *Gentlemen: This is to inform you of my decision to accept the challenging offer of a directorship at the High-Polymer Laboratory of your Vancouver facility. . . .*

Eunice, savior and soul buddy, after juggling several offers chose Du Pont. Other classmates went to Monsanto, Geigy, Dow, Upjohn, American Cyanamid. I was tendered a single offer from Rohm & Margulies—industrial detergents and wetting agents (read *soap factory*)—in Cranston, R.I. With the job went a 2-A deferment, no mean consideration. The Korean War was unfolding, not one of America's best or wisest adventures. Tales of corpses rotting on stark baking hillsides were circulating stateside.

"You'll start out making aryl sulfonates from benzyl mercaptan," said research director Melvin Franks. "I assume you're acquainted with the mercaptan family?"

My stomach hollowed out. "Just in passing . . . not intimately." Dr. Franks looked at me hard and long. The blaze of black eyes and gleam of copper skin against the open-neck khaki shirt conveyed the aspect of a stereotype player in an old Jeff Chandler movie: an indomitable but minor kibbutz character in the shadow of the Golan Heights. The mercaptans—Jesus God, I remembered those sulfur-based mothers from Advanced Organic Synthesis: oily, tenacious, singularly foul liquids.

Over the next four weeks I would become increasingly, reluctantly tight with the mercaptan family—a nasty, unrelenting brood, penetrating the pores, clinging to the skin like cat hair to crinoline. Despite nightly scouring with the harshest of yellow laundry soaps, followed by an expensive pine-scented cake, the funky fragrance trailed after me like the cloud of gloom following Joe Btfsplk, the Al Capp cartoon character with the unpronounceable surname. Empty seats materialized beside me at lunch counters, on buses; half rows became vacant fore and aft in movie theaters. Karen, a freckled blonde from Analytical, a recent Brown graduate, fled my pad on our second date, trampling the amenities: "Sorry, and please don't take it personally, but it'd be like sleeping with a dead mouse." I was beginning to understand why I'd had no competition for this gig.

Dr. Franks called me in for a conference. "I'd like to see more results on the sulfonates. Margulies will be back this weekend, and all I'll have to show him is a month's salary expended on the synthesis and identification of a couple of molecules that a high school senior with no particular scientific bent could have accomplished with a Woolworth set in a basement washtub." The kibbutz veteran had a variety of smiles to go with the sardonic hyperbole, but the smiles were growing less genial of late.

Christ almighty, all this hassle over a system of molecules whose concerted odors would rout a covey of skunks from a garbage dump, whose main functions included the emulsification of insecticides and the cleaning of diapers. My mistake had been panicking in the face of a few rejections, grabbing the first farmclub cracker outfit that vouchsafed an overture. Should have held out for one of the giant impersonal combines, a billet deep in the bowels of Monsanto or Geigy; at least I'd be a member of a team: less emphasis on immediate results, moratorium on anxiety and swamp-water reactants; the prospect of working on something with a shade more social relevance. Yet from Franks's perspective I had to admit that the success rate of my experiments was abysmally low, and whether it was my basic unsuitability for the field or the frigging molecules balking for no logical reason was anybody's guess.

"I think I know where the bottleneck is. I'll—"

"Let me look at your experimental procedures before you leave tonight. From what I read in your reports you're not seeing your analysis through, just whacking off and coming quick."

Whew. Choice rhetoric from a research head and Princeton Ph.D. My options were limited. To jump ship so soon would create problems of professional references, generate a possible fatal interim of unemployment during which my 2-A occupational deferment would evaporate like a beaker of ether on a steam bath. Two decades later kids would dodge the draft, flee the country, or desert the field of battle and be applauded in some quarters. Corporate executives would drop out, choose to deflect coronaries and colitis by driving hacks or teaching at community colleges; M.D.'s would join communes, Ph.D.'s become carpenters or hog farmers in East Dandelion, and they'd be respected for their transgressions. But I was walled into Truman-Eisenhower country, and the ramparts were sturdy.

I badly needed a haven and found it at Club Trocadero, a lively cellar dive on the outskirts of Providence (to which Cranston attaches like a tick to a mangy Great Dane). I hung out there a couple of nights a week, sitting in with a very decent piano-guitar-bass trio playing old gold from the Ellington-Arlen-Gershwin songbooks. I had found an industrial-strength cologne to mask the pungent mercaptans and now reeked of mint. Ladies passing the piano smiled and fanned themselves ("Puh-leeze") and the bartender told me

I smelled like "a goddamn fruit." (It's mill-town New England, folks, 1951; I didn't take it personally.) But I was making music in lieu of soap, and the joint was jumping!

TWO GRINDING HOURS spent trying to replace one infinitesimal, intractable, *invisible* hydrogen atom with an ethyl group, consoling myself with reveries of struggles on a grander scale: polio vaccine, Manhattan Project (as Oppenheimer's indispensable assistant and trusted confidant). If only I could *see* the connective carbon-carbon bonds, some manner of optical device clamped to the face like a scuba diver's mask, which would render the vital ligaments visible. . . . Undigested reactants sat in the three-necked flask like a revolting bowl of weekold bean soup, various unseemly components layering out. Miserable no-count molecules just lying there, crapped out, like winos in the library shade. I felt the immemorial urge, as elemental as hunger: a couple of swift kicks to the counter, loosen the gears—*unh! unh!* An icy tingle at the back of my neck, a half-sensed shadow in the doorway. Who lurks?

"What gives?" Over the shoulder I glimpsed Franks smiling one of his indecipherable smiles, staring at the scuffed shoe encasing the foot assaulting the base of the counter.

I faced full around, thoughts whirling, formulating hopeless, untenable explanations. A nervous elbow jarred the dropping funnel feeding alkylating agent through the flask's middle neck. The turncock loosened, fell, smashed on the counter, simultaneously releasing a stinking stream of ethyl bromide.

"Shoot. . . . Sorry, Doctor."

"In again, out again, Hooligan."

Grabbing a beaker and a scrap of cheesecloth, I attempted to collect the abominable liquid dripping off the soapstone, watching coolly (with scientific detachment, if you will) the hand holding the beaker tremble like an old man's. With so many people loathing or laboring vainly in their chosen professions you'd think there'd be thousands running amok in the corridors, tearing at one another's throat.

"The stuff goes for two eighty-nine a liter. And it shouldn't smell that rank if you've been double distilling." Gnawing a corner of his lip, uncharacteristically contemplative, deep-black eyes gleaming like pitchstone. "One of the plant workmen said he saw you tinkling the piano at a saloon on Broad Street the other night. Beer bottles were lined up on top like clay ducks, he said."

I set the stinking beaker carefully on the counter, the arrow at my heart. "The working day was over. The beer bottles weren't all mine."

"If I've got the right joint it's crawling with two-dollar whores. Maybe that's why you've been dragging your ass around here like a zombie."

A lot of things drifted through my mind. Bivouacking in Pusan, my proud mama and hard-toiling brother, my fragile father in his grave. Classmates and cousins preparing for medicine, law, engineering. Not a whorehouse piano player in the lot.

Taped high on the far wall of the lab was a sign presumably appropriated from a roadside dump, the whimsical handiwork of a previous tenant: DO NOT DUMP GARBAGE, REFUSE. (The scintilla of humor resides in pronouncing the last word as a verb.) In the waning gray afternoon light of this reeking lab in Hellhole, R.I., under the anthracite gaze of my superior, the sign seemed to resonate, to send me an irresistible message. With an audible sigh I peeled off my grungy lab coat of many colors and tossed it on the counter.

There's a shock that comes with the recognition that your talent in a field you've spent four years preparing for is irremediably minor, and with the shock comes the impulse to deflect it with whatever antic or frivolous weapon is at hand.

"Well, I tried. Soap was never my bag. I'm awful tired of the mercaptan family and tired of stinking up the whole state of Rhode Island, so I'm going to split. But I'm no jack-off artist."

"If that's your decision, I can live with it," Franks said. *Meshuggener.*

THE FOLLOWING AFTERNOON, in one of the cubbyhole offices of the Paramount Agency for Professional Personnel in Boston—the third such agency visited that day—I faced a sallow red-haired man in a shortsleeve white shirt and rep tie; the brass-on-wood nameplate read, WM. MARIGOLD, JR.

"B.A. Organic chemistry, Cornell. Shouldn't be too hard to place." He flipped over the application. "You've put down 2-A for marital status."

"Marital . . ."

"Not martial."

"Ah—Single. MacArthur slip." The chills were beginning to take over.

"Rohm and Margulies . . . work not suited. . . . What kind of work was this?"

"Synthetic detergents and wetting agents. Basically we made soap."

"Doesn't sound terribly stimulating."

"Dull as dishwater." A wry drop of levity wrung from a shriveled heart.

Oblivious, Marigold riffled through a card file; his narrow wedge-shaped face made me think of an arrowhead I had uncovered as a kid on the muddy bank of Lake Quinsigamond. "American Cyanamid has two openings for recent graduates—involves a six-month training program. They're in Providence."

The fetid aroma of mercaptan hung in the heated air. "I'd prefer not to return so soon to that locale."

"Union Carbide—two years' experience. Would you be interested in something in New Bedford?"

"New Bedford . . ."

"Seacol Company. They make various agents from sea plants. Involves some knowledge of polymers and hydrocolloids, experience unnecessary. Pay is quite good—four hundred and fifty to four seventy-five, depending on qualifications."

I ransacked my muddled memory. "I did take a course in colloid chemistry. And a senior-year seminar in polymers . . ."

"Are you interested?"

"Would this, ah, involve a 2-A deferment?"

"That depends on the individual employer and community conditions. In this line of work it's always a likelihood. If you're interested I'll get right on it."

"Interested."

Marigold lifted the phone, dialing without taking his eyes from the file card, and remarked flatly while waiting out the connection, "I have a son at Fort Ord shipping out next week."

He spoke for a minute or two, mostly reading from my application, then covered the mouthpiece. "Can you go to New Bedford for an interview tomorrow afternoon?"

New Bedford. Not too bad a haul, a lot closer than Pusan. I nodded, the activity of least resistance.

I departed Paramount in a woolly funk. Sea *plants,* the man said. Was that the same as sea*weed?* Kelp, algae, the polysaccharides—I began to remember. Nothing about this gig felt right, not my forte at all, as they say in the music biz. New Bedford: all that came to mind was *Down to the Sea in Ships,* Richard Widmark; mudflats, gulls, tall masts in the mist. I spent the rest of the afternoon and early evening in the Boston Public Library wading through *Natural Plant Hydrocolloids, Polysaccharide Chemistry,* and *Seaweed at Ebb Tide.*

Sleepless, sweat-bathed on an iron-frame bed in a fleabag hotel, I thought of phoning Herbie in Corning, getting his take on the situation. I hadn't yet told him (or my mother) about the fiasco in Cranston. I thought I knew what he'd say. His breath would kind of explode on the phone the way it does when he's stressed or thinking of a next move: "I don't see that you have any choice, kid.

You don't want to be futzing around looking for work with the draft board breathing on your neck. The last I heard they weren't passing out 2-A's to saloon ivory tinklers, so bite the bullet, bozo. Welcome to the big bad world." *Of seaweed?* I watched traffic lights drift across the fissured ceiling, thoughts tangling in windrows of kelp and its gelatinous extracts: algin, agar, carrageenin, floridean starch. . . . Suddenly I knew I'd pass that interview tomorrow (contingent on Franks, if contacted for a reference, having the heretofore unobserved decency and compassion to remain noncommittal). Two-A status secured, I'd accept the offer, find a room in a shingled weatherbeaten house facing the waterfront, which would smell of decaying bivalves and gull guano. The prospect of a return session with Marigold or counterpart, more phone calls, interviews, libraries, fleabags, filled me with dread. I was being transported willy-nilly to the southeastern coast of Massachusetts, boneless, unresisting, will paralyzed by chloral hydrate or one of the more insidious psychic drugs.

IF YOU LOOK CLOSELY—there, halfway to the domed white ceiling of the cavernous, shuddering plant—you can spot a forlorn figure, the former apprentice pants cleaner and Rohm & Margulies screw-up bizarrely attired in rubber boots and long flapping khaki coat. Climbing a slender metal ladder affixed to the side of a cylindrical extraction tank the size of a Great Plains water tower. Carrying up with him stopwatch, clipboard, tubes of litmus paper, and three-foot-long thermometer. At this dizzying height the biosphere is heavy with a wild rank odor of marshland and rotting fish, and for the second time in three months he begins to understand why there has been little competition for the gig. Other figures are perched on scaffolding surrounding similar tanks, the great dank enclosure vibrating with the din of rotary dryers, pressure filters, centrifuges. (He senses the plant director, stationed somewhere below in the wet bowels of the building, watching him.) His eyes have a haunted look—he has slept very little, wading through swampy and arcane texts into the wee hours—and his limbs feel as if they are attached to invisible wires as he mechanically plunges the thermometer into the roiling tank awash with ungainly skeins of scarlet seaweed, tests with litmus, and jots down figures on the clipboard. Brackish steam rises in his face, but his hands are raw with cold. If the Loch Ness monster had reared its slithery head from the seaweed he would not have been startled; nor would the creature's emergent thrashing have had any effect on his data.

When the boxes on the clipboard sheet are all filled he descends the ladder. The cement floor, fairgrounds-broad and sloping toward numerous drains,

still manages to collect sizable pools of brine. He has noticed that no one bothers to circumvent the pools; all take the direct transatlantic route, sloshing
straight through in their boots (though without any visible attendant childhood pleasure), as does the plant manager now, looking up from consultation
at the base of a sedimentation tank to catch him standing idle and lost at the
foot of the ladder. The director's soft uninflected voice advises, as he scans the
clipboard, "I'm afraid these figures won't be very useful. Your time intervals are
too erratic." He deigns no answer; a response could only be obtuse or jocular.
Cockamamie tank might as well have been stocked with sea horses and blue-
point oysters. The machinery's throbbing grows distant now, as if beachhouse
storm windows have been closed on a pounding surf. He feels the director's
hand on his shoulder. ". . . bound to be confusing the first week or so. We'll let
this ride for now. I suggest you spend the rest of the day in the library reac-
quainting yourself with the packet of materials you have. Give particular focus
to the thixotropic-flow charts."

Thixotropic flow? That one wasn't in the Boston stacks.

Gentle monarch of the sea still touching his shoulder (it is in part the gen-
tleness that will undo him), regarding him with a slight frown; thick lenses
lend his eyes a distorted, silverish, fishy cast. "Do you remember where the library is?"

"Not really."

"Second landing behind centrifuge D-4, southwest corner."

"Can I borrow your compass?" In-again, out-again Hooligan turning tail
with a cunning smile that breaks into introspective laughter as he heads
straight for the first magnum-size pool and splashes on through, savoring the
frolicking spray, shedding all burden and responsibility. *Hey, this is fun. . . .*

EVERY FAMILY, it has been said, is entitled to one nervous breakdown. All
indicators pointed this way. I didn't disappoint; the psychological chickens of
a loopy childhood and later bad choices came home to roost. Supine and
mostly mute day in, day out (*not a bad tune*) through endless indistinguishable
weeks, blankets pulled up to a white tip of nose. Anxious mama peering down,
trying to wheedle an explanation from a post, doubtless contemplating the
malign paternal legacy—manifest in the trembling smile, dark-circled eyes
flooded with apprehension and rue; you could drown in those eyes. Even in his
detached state the kid could tell you something about collapse, deserting the
field, whatever name you care to put on it. It's unrelieved stupor, vile brass
coating the tongue, an infant's involuntarily clenched fingers and toes; the least

goddamn thing irritates. It's night sweats; it's sodium's cemetery light at dawn that ices the bones, plants invisible spiders on the skin, sleet in the heart.

"What the hell's going on?" Dark-suited big bro, summoned reluctantly from Corning, filling the doorway like a fireplug.

"As you can plainly see, not a blessed thing." Herbie's solid, comforting presence spurring him to arch articulation.

"Pull the blanket off your face, I can't hear."

"They wanted me to dump garbage. I refused."

"I take off from work, fly five hundred miles, and you talk gibberish to me."

"What basically happened—I got drowned in a vat of seaweed."

"Will you for Christ sake get out of bed? You're scaring Ma and making me sick."

"Up yours."

"Great. This I fly across two states for."

Dr. Harry Fine, old-line house-call family sawbones arrives, is brought up to date and makes a cursory physical examination.

"What I believe we're looking at here is a temporary failing of courage shading into a mild breakdown. Which can occur with sensitive individuals"—*sensitive* accorded a disdainful equal stressing of all syllables—"when we cut too severely against the grain or batter our heads too often against an unforgiving wall."

"You're sure it's nothing more? He just lies there day after day. Whatever I do seems—Harry, I can't get through to him." Stricken mama clutching her hands, the fear in her eyes tearing at him.

"Sophie, I know this one from way back. I delivered him, remember? What's going on here has nothing to do with his father's problems, so put that out of your head."

They're talking about him as if he's not there at all—and it's true, he's not all there but is experiencing the first flush of shame and restlessness that will eventually uncurl the baby toes and fingers, brush the daddy longlegs from the skin. Later will come self-revulsion and remorse for putting Sophie and Herbie through this wringer.

"He might be faking it, the draft board's after him," Herbie puts in, recounting the substance of a recent bureaucratic communiqué—*According to our information your employment with Seacol Company terminated on October 21. You are required by law to notify your local draft board within ten days of any change of employment. Please contact this board immediately to schedule a pre-introduction physical examination*—leaving the beleaguered kid, as if he hasn't enough crap to contend with, envisioning his corpse exposed on a frozen hillside in a dumb war 10,000 miles from home that no one in his right mind gives a good goddamn about.

"Draft board," Dr. Harry snorts. "Your brother's temporarily a basket case. He'd be as much use to Uncle Sam as a fly swatter against a panzer division."

The next morning, perceiving his usefulness at an end, Herbie prepared to depart. "Enough's enough, okay?" He hung in the doorway, ill at ease. "Get up and make your bed, give Ma a break. I'll be in touch, you know, if you need me."

"Tubbo." A way-back nickname unused for fifteen years. Half out the door, he turned warily. "What the hell is thixotropic flow?"

Tubbo started to reply and broke off, staring in consternation at the tears swarming in kid brother's eyes.

A MONTH LATER, having predictably failed my pre-induction physical (*Kind of shaky there, aren't you, son? What's this about? Fear or something else?*), I took off like a singed cat for the Lake Quinsigamond dives on the Worcester-Shrewsbury line and found refuge at the Jolly Roger. It was a dump, but a convivial dump: the piano a jangling drink-stained relic, the decor bare-bones nautical (tattered fish netting slung from ceiling beams, hurricane lamps and seashell ashtrays, bogus portholes), the pay sweatshop ($4/night), the clientele working class in your face ("When my chick gets back from the toilet, start 'You Are My Sunshine' and keep playin' it till I signal you"). The family didn't begrudge me the lowlife gig. They were happy to see me back on my feet, however unsteadily, pulling it together. The shame was still fresh; they'd seen me raw and mewling, helpless as an infant. The interim employment, they felt, would be no more permanent than an Ace bandage. They were wrong.

From my lakeside sanctuary I cranked out the tunes of the day—"Autumn Leaves," "A Kiss to Build a Dream On," "While We're Young"—breathing hungrily the healing balm of smoke, cheap rye, and spilled brew (displacing the fading reek of kelp and rotting fish); swinging like a creaky garden gate but going with the grain, willing myself to forgetfulness. The dinky ruins of a scientific career that barely got off the ground strewn like seaweed at ebb tide seventy miles south in a barnacle-encrusted whaling port. Let us say *Kaddish*.

NATALIE ANGIER

Furs for Evening, but Cloth Was the Stone Age Standby

FROM THE *NEW YORK TIMES*

It's hard to dislodge notions of cavepeople in unchic pelts. Leave it, then, to the popular and always entertaining science journalist Natalie Angier to find a group of researchers who have discovered that prehistoric humans were not without fashion sense.

Ah, the poor Stone Age woman of our kitschy imagination. When she isn't getting bonked over the head with a club and dragged across the cave floor by her matted hair, she's hunched over a fire, poking at a roasting mammoth thigh while her husband retreats to his cave studio to immortalize the mammoth hunt in fresco. Or she's Raquel Welch, saber-toothed sex kitten, or Wilma Flintstone, the original Roccer Mom. But whatever her form, her garb is the same: some sort of animal pelt, cut nasty, brutish and short.

Now, according to three anthropologists, it is time to toss such hidebound cliches of Paleolithic woman on the midden heap of prehistory. In a new analysis of the renowned "Venus" figurines, the hand-size statuettes of female bodies carved from 27,000 to 20,000 years ago, the researchers have found evidence that the women of the so-called upper Paleolithic era were far more accomplished, economically powerful and sartorially gifted than previously believed.

As the researchers see it, subtle but intricate details on a number of the figurines offer the most compelling evidence yet that Paleolithic women had already mastered a revolutionary skill long thought to have arisen much later in
human history: the ability to weave plant fibers into cloth, rope, nets and
baskets.

And with a flair for textile production came a novel approach to adorning
and flaunting the human form. Far from being restricted to a wardrobe of
what Dr. Olga Soffer, one of the researchers, calls "smelly animal hides," Paleolithic people knew how to create fine fabrics that very likely resembled
linen. They designed string skirts, slung low on the hips or belted up on the
waist, which artfully revealed at least as much as they concealed. They wove
elaborate caps and snoods for the head, and bandeaux for the chest—a series
of straps that amounted to a cupless brassiere.

"Some of the textiles they had must have been incredibly fine, comparable
to something from Donna Karan or Calvin Klein," said Dr. Soffer, an archae-
ologist with the University of Illinois in Urbana-Champaign.

Archaeologists and anthropologists have long been fascinated by the Venus
figurines and have theorized endlessly about their origin and purpose. But
nearly all of that speculation has centered on the exaggerated body parts of
some of the figurines: the huge breasts, the bulging thighs and bellies, the well-
defined vulvas. Hence, researchers have suggested that the figurines were fer-
tility fetishes, or prehistoric erotica, or gynecology primers.

"Because they have emotionally charged thingies like breasts and buttocks,
the Venus figurines have been the subject of more spilled ink than anything I
know of," Dr. Soffer said. "There are as many opinions on them as there are
people in the field."

In their new report, which will be published in the spring in the journal
Current Anthropology, Dr. Soffer and her colleagues, Dr. James M. Adovasio
and Dr. David C. Hyland of the Mercyhurst Archaeological Institute at Mercyhurst College in Erie, Pa., point out that voluptuous body parts notwithstanding, a number of the figurines are shown wearing items of clothing. And when
they zeroed in on the details of those carved garments, the researchers saw
proof of considerable textile craftsmanship, an intimate knowledge of how
fabric is woven.

"Scholars have been looking at these things for years, but unfortunately,
their minds have been elsewhere," Dr. Adovasio said. "Most of them didn't recognize the clothing as clothing. If they noticed anything at all, they misinterpreted what they saw, writing off the bandeaux, for example, as tattoos or
body art."

Scrutinizing the famed Venus of Willendorf, for example, which was discovered in lower Austria in 1908, the researchers paid particular attention to the statuette's head. The Venus has no face to speak of, but detailed coils surround its scalp. Most scholars have interpreted the coils as a kind of paleo-coiffure, but Dr. Adovasio, an authority on textiles and basketry, recognized the plaiting as what he called a "radially sewn piece of headgear with vertical stem stitches."

Willendorf's haberdashery "might have looked like one of those woven hats you see on Jamaicans on the streets of New York," he said, adding, "These were cool things."

On the Venus of Lespugue, an approximately 25,000-year-old figurine from southwestern France, the anthropologists noticed a "remarkable" degree of detail lavished on the rendering of a string skirt, with the tightness and angle of each individual twist of the fibers carefully delineated. The skirt is attached to a low-slung hip belt and tapers in the back to a tail, the edges of its hem deliberately frayed.

"That skirt is to die for," said Dr. Soffer, who, before she turned to archaeology, was in the fashion business. "Though maybe it's an acquired taste."

To get an idea of what such an outfit might have looked like, she said, imagine a hula dancer wrapping a 1930's-style beaded curtain around her waist. "We're not talking protection from the elements here," Dr. Soffer said. "This would have been ritual wear, if it was worn at all, a way of communicating with higher powers."

Other anthropologists point out that string skirts, which appear in Bronze-Age artifacts and are mentioned by Homer, may have been worn at the equivalent of a debutantes ball, to advertise a girl's coming of age. In some parts of Eastern Europe, the skirts still survive as lacy elements of folk costumes.

The researchers presented their results earlier this month at a meeting on the importance of perishables in prehistory that was held at the University of Florida in Gainesville. "One of the most common reactions we heard was, 'How could we have missed that stuff all these years?' " Dr. Adovasio said.

Dr. Margaret W. Conkey, a professor of anthropology at the University of California at Berkeley, and co-editor, with Joan Gero, of *Engendering Archaeology* (Blackwell Publishers, 1991) said, "They're helping us to look at old materials in new ways, to which I say bravo!"

Not all scholars had been blinded by the Venutian morphology. Dr. Elizabeth Wayland Barber, a professor of archaeology and linguistics at Occidental College in Los Angeles, included in her 1991 volume *Prehistoric Textiles* a chapter arguing that some Venus figurines were wearing string skirts. The recent work from Dr. Soffer and her colleagues extends and amplifies Dr. Barber's original observations.

The new work also underscores the often neglected importance of what Dr. Barber has termed the "string revolution." Archaeologists have long emphasized the invention of stone and metal tools in furthering the evolution of human culture. Even the names given to various periods in human history and prehistory are based on heavyweight tools: the word "Paleolithic"—the period extending from about 750,000 years ago to 15,000 years ago—essentially means "Old Stone Age." And duly thudding and clanking after the Paleolithic period were the Mesolithic and Neolithic, or Middle and New Stone Age, the Bronze Age, the Iron Age, the Industrial Age.

But at least as central to the course of human affairs as the invention of stone tools was the realization that plant products could be exploited for purposes other than eating. The fact that some of the Venus figurines are shown wearing string skirts, said Dr. Barber, "means that the people who made them must also have known how to make twisted string."

With the invention of string and the power to weave, people could construct elaborate yet lightweight containers in which to carry, store and cook food. They could fashion baby slings to secure an infant snugly against its mother's body, thereby freeing up the woman to work and wander. They could braid nets, the better to catch prey animals without the risk of hand-to-tooth combat. They could lash together wooden logs or planks to build a boat.

"The string revolution was a profound event in human history," Dr. Adovasio said. "When people started to fool around with plants and plant byproducts, that opened vast new avenues of human progress."

In the new report, the researchers argue that women are likely to have been the primary weavers and textile experts of prehistory, and may have even initiated the string revolution in the first place—although men undoubtedly did their share of weaving when it came to making hunting and fishing nets, for example. They base that conclusion on modern crosscultural studies, which have found that women constitute the great bulk of the world's weavers, basketry makers and all-round mistresses of plant goods.

But while vast changes in manufacturing took the luster off the textile business long ago, with the result that such "women's work" is now accorded low status and sweatshop wages, the researchers argue that weaving and other forms of fiber craft once commanded great prestige. By their estimate, the detailing of the stitches shown on some of the Venus figurines was intended to flaunt the value and beauty of the original spinsters' skills. Why else would anybody have bothered etching the stitchery in a permanent medium, if not to boast, whoa! Check out these wefts!

"It's made immortal in stone," Dr. Soffer said. "You don't carve something like this unless it's very important."

The detailing of the Venutian garb also raises the intriguing possibility that the famed little sculptures, which rank right up there with the Lascaux cave paintings in the pantheon of Western art, were hewn by women—moonlighting seamstresses, to be precise. "It's always assumed that the carvers were men, a bunch of guys sitting around making their zaftig Barbie dolls," Dr. Soffer said. "But maybe that wasn't the case, or not always the case. With some of these figurines, the person carving them clearly knew weaving. So either that person was a weaver herself, or he was living with her. He's got an adviser."

Durable though the Venus figurines are, Dr. Adovasio and his co-workers are far more interested in what their carved detailing says about the role of perishables in prehistory. "The vast bulk of what humans made was made in media that hasn't survived," Dr. Adovasio said. Experts estimate the ratio of perishable objects to durable objects generated in the average culture is about 20 to 1.

"We're reconstructing the past based on 5 percent of what was used," Dr. Soffer said.

Because many of the items that have endured over the millennia are things like arrowheads and spear points, archaeologists studying the Paleolithic era have generally focused on the ways and means of that noble savage, aka Man the Hunter, to the exclusion of other members of the tribe.

"To this day, in Paleolithic studies we hear about Man the Hunter doing such boldly wonderful things as thrusting spears into woolly mammoths, or battling it out with other men," Dr. Adovasio said. "We've emphasized the activities of a small segment of the population—healthy young men—at the total absence of females, old people of either sex and children. We've glorified one aspect of Paleolithic life ways at the expense of all the other things that made that life way successful."

Textiles are particularly fleeting. The oldest examples of fabric yet discovered are some carbonate-encrusted swatches from France that are about 18,000 years old, while pieces of cordage and string dating back 19,000 years have been unearthed in the Near East, many thousands of years after the string and textile revolution began.

In an effort to study ancient textiles in the absence of textiles, Dr. Soffer, Dr. Adovasio and Dr. Hyland have sought indirect signs of textile manufacture. They have pored over thousands of ancient fragments of fired and un-fired clay, and have found impressions of early textiles on a number of them, the oldest dating to 29,000 B.C. But the researchers believe that textile manu-facture far predates this time period, for the sophistication of the stitchery rules out its being, as Dr. Soffer put it, "what you take home from Crafts 101." Dr. Adovasio estimates that weaving and cord-making probably go back to the year 40,000 B.C. "at a minimum," and possibly much further.

Long before people had settled down into towns with domesticated plants and animals, then, while they were still foragers and wanderers, they had, in a sense, tamed nature. The likeliest sort of plants from which they extracted fibers were nettles. "Nettle in folk tales and mythology is said to have magic properties," Dr. Soffer said. "In one story by the Brothers Grimm, a girl whose two brothers have been turned into swans has to weave them nettle shirts by midnight to make them human again." The nettles stung her fingers, but she kept on weaving.

But what didn't make it into Grimms' was that when the girl was done with the shirts, she took out a chisel, and carved herself a Venus figurine.

STEPHEN JAY GOULD

A Division of Worms

FROM *NATURAL HISTORY*

If known at all, the name of Lamarck is associated with the theory he cham-pioned and which has since been completely discredited: that acquired traits can be passed along through heredity. This one mistake has overshadowed the many valid achievements of Lamarck's career. In this exciting re-evaluation, the indefatigable Stephen Jay Gould—biologist, evolutionary theorist, and author many times over—rescues Lamarck from the margins of history, re-vealing a subtlety to his thought that has too long been overlooked.

I. The Making and Breaking of a Reputation

On the twenty-first day of the auspiciously named month of Floréal (flowering), in the spring of the year 8 on the French revolutionary calendar (1800 to the rest of the Western world), the former Cheva-lier (knight) but now Citoyen (citizen) Lamarck delivered the opening lecture for his annual course on zoology at the Muséum d'Histoire Naturelle in Paris—and changed the science of biology forever by presenting the first pub-lic account of his theory of evolution. Lamarck then published this short dis-course in 1801, as the first part of his treatise on invertebrate animals, *Système des animaux sans vertèbres* (System of invertebrate animals).

Jean-Baptiste Lamarck (1744–1829) had enjoyed a distinguished career in botany when, just short of his fiftieth birthday, he became a professor of "insects, worms, and microscopic animals" at the Muséum d'Histoire Naturelle, newly constituted by the Revolutionary government in 1793. Lamarck would later coin the term *invertebrate* for his assigned organisms. (In 1802 he also introduced the word *biology* for the entire discipline.) But his original title followed Linnaeus's designation of all spineless animals as either insects or worms, a Procrustean scheme that Lamarck would soon alter. Lamarck had been an avid shell collector and student of mollusks (then classified within Linnaeus's large and heterogeneous category of Vermes, or worms)—qualifications deemed sufficient for his shift from botany.

Lamarck fully repaid the confidence invested in his general biological abilities by publishing distinguished works in the taxonomy of invertebrates throughout the remainder of his career, culminating in the seven volumes of his comprehensive *Histoire naturelle des animaux sans vertèbres* (Natural history of invertebrate animals), published between 1815 and 1822. At the same time, he constantly refined and expanded his evolutionary views, extending his introductory discourse of 1800 into a full book in 1802 (*Recherches sur l'organisation des corps vivants,* or "Research on the organization of living beings"); then into his magnum opus and most famous work in 1809, the two-volume *Philosophie zoologique* (Zoological philosophy); and finally into a statement for the long opening section, published in 1815, of his great treatise on invertebrates.

The outlines of such a career might seem to imply continuing growth of prestige, from an initial flowering to the full bloom of celebrated seniority. But Lamarck's reputation suffered a spectacular collapse, even during his own lifetime, and he died lonely, blind, and impoverished. The complex reasons for his reverses include the usual panoply of changing fashions, powerful enemies, and self-inflicted wounds based on flaws of character (in his case, primarily an overweening self assurance that could not face, or even recognize, the weaknesses in some of his own arguments or the skills of his adversaries). Most prominently, his favored style of science—the construction of grand and comprehensive theories, following an approach that the French call *l'esprit de système* (the spirit of system building), and not always well supported by testable data—became notoriously unpopular after the rise of a hard-nosed, empiricist ethos in early nineteenth-century geology and natural history.

In one of the great injustices of our conventional history, Lamarck's disfavor has persisted to our times, and most people still know him only as the foil to Charles Darwin's greatness—as the man who proposed a silly theory about giraffes stretching their necks to reach high leaves and then passing the fruits of their efforts to their offspring by "inheritance of acquired characters," otherwise known as the hypothesis of use and disuse, in contrast with Darwin's proper theory of natural selection and survival of the fittest.

Indeed, the usually genial Darwin had few kind words for his French predecessor. In letters to his friends, Darwin dismissed Lamarck as an idle speculator with a nonsensical theory. In 1844 he wrote to the botanist Joseph D. Hooker on the dearth of evolutionary thinking (before his own ideas about natural selection): "With respect to books on the subject, I do not know of any systematical ones except Lamarck's, which is veritable rubbish." To his guru, the geologist Charles Lyell (who had accurately described Lamarck's system for English readers in the second volume of his *Principles of Geology*, published in 1832), Darwin wrote in 1859, just after publishing the *Origin of Species:* "You often allude to Lamarck's work; I do not know what you think about it, but it appeared to me extremely poor; I got not a fact or idea from it."

But these later and private statements did Lamarck no practical ill. Far more harmfully, and virtually setting an "official" judgment from that time forward, his eminent colleague Georges Cuvier (the brilliant biologist, savvy statesman, distinguished man of letters, and Lamarck's younger and anti-evolutionary fellow professor at the museum) used his established role as writer of *éloges* (obituary notices) for deceased colleagues to compose a cruel masterpiece in the genre of "damning with faint praise"—a document that fixed and destroyed Lamarck's reputation. Cuvier began with cloying praise and then described his need to criticize as a sad necessity:

> In sketching the life of one of our most celebrated naturalists, we have conceived it to be our duty, while bestowing the commendation they deserve on the great and useful works which science owes to him, likewise to give prominence to such of his productions in which too great indulgence of a lively imagination had led to results of a more questionable kind, and to indicate, as far as we can, the cause or, if it may be so expressed, the genealogy of his deviations.

Cuvier then proceeded to downplay Lamarck's considerable contributions to anatomy and taxonomy and to excoriate his senior colleague for fatuous speculation about the comprehensive nature of reality. Cuvier especially ridiculed his subject by contrasting his caricature of Lamarck's ideas with the sober approach of proper empiricism:

These [evolutionary] principles once admitted, it will easily be perceived that nothing is wanting but time and circumstances to enable a monad or a polypus gradually and indifferently to transform themselves into a frog, a stork, or an elephant. . . . A system established on such foundations may amuse the imagination of a poet; a metaphysician may derive from it an entirely new series of systems; but it cannot for a moment bear the examination of anyone who has dissected a hand . . . or even a feather.

Cuvier's *éloge* reeks of exaggeration and unjust ridicule, especially toward a colleague ineluctably denied the right of response—the reason, after all, for our venerable motto *De mortuis nil nisi bonum* (Say only good of the dead). But Cuvier did base his disdain on a legitimate substrate, for Lamarck's writing certainly shows a tendency to grandiosity in its comprehensive pronouncements, combined with frequent refusal to honor, or even to consider, alternative views with strong empirical support.

L'esprit de système, the propensity for constructing complete and overarching explanations based on general and exceptionless principles, may apply to some corners of reality but works especially poorly in the maximally complex world of natural history. Lamarck did feel drawn to this style of system building, and he showed no eagerness to acknowledge exceptions or to change his guiding precepts. But the rigid and dogmatic Lamarck of Cuvier's caricature can be regarded only as a great injustice, for the man himself did maintain appropriate flexibility before nature's richness and did eventually alter the central premises of his theory when his own data on the anatomy of invertebrate animals could no longer sustain his original view.

This fundamental change—from a linear to a branching system of classification for the basic groups, or phyla, of animals—has been well documented in standard sources of modern scholarship about Lamarck (principally in Richard W. Burkhardt Jr.'s *The Spirit of System: Lamarck and Evolutionary Biology*, Harvard University Press, 1977; and Pietro Corsi's *The Age of Lamarck: Evolutionary Theories in France 1790–1830*, University of California Press, 1988). But the story of Lamarck's intellectual journey remains incomplete, for both the first explicit statement and the final conclusion have been missing from the record—the beginning because Lamarck noted his first insight as a handwritten insertion, heretofore unpublished, in his own copy of his first printed statement about evolution (the Floréal address of 1800, recycled as the preface to his 1801 book on invertebrate anatomy); and the ending because his final book of 1820, *Système analytique des connaissances positives de l'homme* (Analytic system of positive knowledge of man), has been viewed only as an obscure swan song about psychology—a rare book even more rarely consulted, despite a fascinating section containing a crucial and novel wrinkle on Lamarck's continually changing views about the classification of animals. Stories deprived of both beginnings and endings cannot satisfy our urges for fullness or completion—and I am grateful for this opportunity to supply these terminal anchors in this essay.

II. Lamarck's Theory and Our Misreadings

LAMARCK'S ORIGINAL EVOLUTIONARY SYSTEM—the logical, pure, and exceptionless scheme that nature's intransigent complexity later forced him to abandon—featured a division of causes into two independent sets, responsible for progress and diversity, respectively. (Scholars generally refer to this model as the "two factor theory.") On the one hand, the "force that tends incessantly to complicate organization" (*la force qui tend sans cesse à composer l'organisation*) leads evolution inexorably upward, beginning with the spontaneous generation of infusorians (single-celled animals) from chemical precursors, and moving on toward human intelligence.

But Lamarck recognized, on the other hand, that the riotous diversity of living organisms could not be ordered into a neat and simple series of linear advances—for what could come directly before or after such marvels of adaptation as long-necked giraffes, moles without eyes, flatfishes with both eyes on one side of the body, snakes with forked tongues, or birds with webbed feet? Lamarck therefore advocated linearity only for the "principal masses," or major anatomical designs of life's basic phyla. Thus, he envisioned a linear sequence—mounting in perfect, progressive regularity—from infusorian to jellyfish to worm to insect to mollusk to vertebrate. He then depicted the special adaptations of particular lineages as lateral deviations from this main sequence.

These special adaptations originate by the second set of causes, labeled by Lamarck as "the influence of circumstances" (*l'influence des circonstances*). Ironically, this second (and subsidiary) set has descended through history to become the Lamarckism of modern textbooks and anti-Darwinian iconoclasts, while the more important first set of linearizing forces has been forgotten. This second set—based on change of habits as a spur to adaptation in new environmental circumstances—invokes the familiar (and false) doctrines now called Lamarckism: the inheritance of acquired characters and the principle of use and disuse.

Lamarck invented nothing original in citing these principles of inheritance, for both doctrines represented the folk wisdom of his time (despite their later disproof in the new world of Darwin and Gregor Mendel). Thus, the giraffe stretches its neck throughout life to reach higher leaves on acacia trees, and the shorebird extends its legs to remain above the rising waters. This sustained effort leads to longer necks or legs—and these rewards of hard work then descend to offspring in the form of altered heredity (the inheritance of acquired characters, either enhanced by use, as in these cases, or lost by disuse, as in eyeless moles or blind fishes living in perpetually dark caves).

As another irony and injustice (though abetted, in part, by Lamarck's own unclear statements), the ridicule surrounding Lamarck's theory, ever since Cuvier's *éloge* and Darwin's dismissal, has always centered on the charge that his views represent a sad throwback to the mystical vitalism of bad old times, before modern science enshrined testable mechanical causes as the proper source of explanation. What genuine understanding, critics charge, can possibly arise from claims about vague and unknowable powers inherent in life itself, either propelling organisms upward by an intrinsic complexifying force (recalling Molière's famous mock of vitalistic medicine, exemplified in the claim that morphine induces sleep *quia est in eo virtus dormativa,* "because it contains a dormitive virtue"), or sideward by some ineffable "willing" to build an adaptive branch by sheer organic effort or desire?

In a famous letter to Hooker, his closest confidant, Darwin first admitted his evolutionary beliefs in 1844 by contrasting his own mechanistic account with a caricature of Lamarck's theory: "I am almost convinced . . . that species are not (it is like confessing a murder) immutable. Heaven forfend me from Lamarck nonsense of a 'tendency to progression,' 'adaptations from the slow willing of animals,' etc.! But the conclusions I am led to are not widely different from his; though the means of change are wholly so." And Cuvier, in a public forum, ridiculed Lamarck's second set of adaptive forces in the same tone: "Wants and desires, produced by circumstances, will lead to other efforts, which will produce other organs. . . . It is the desire and the attempt to swim that produces membranes in the feet of aquatic birds. Wading in the water . . . has lengthened the legs of such as frequent the sides of rivers."

Lamarck hurt his own cause by careless and easily misinterpreted statements. His talk about an "inner urge" (*sentiment intérieur*) to propel the upward force, or about organisms obeying "felt needs" (*besoins,* in his terminology) to induce sideward branches of adaptation, led to suspicions about mysterious and unprovable vitalistic forces. But, in fact, Lamarck remained a dedicated and vociferous materialist all his life—a credo that surely represents the most invariable and insistent claim in all his writings. He constantly sought to devise mechanical explanations, based on the physics and chemistry of matter in motion, to propel both sets of forces—linear and lateral. I do not claim that his efforts were crowned with conspicuous success—particularly in his speculative efforts to explain the linear sequence of animal phyla by positing an ever more vigorous and ramifying flow of fluids that carved out spaces for organs, and channels for blood, in progressively more complex bodies. But one cannot deny his consistent conviction. *"La vie . . . n'est autre chose qu'un phénomène physique* (Life is nothing but a physical phenomenon)," he wrote in his last book of 1820. In a famous article, written to rehabilitate Lamarck on the occasion of the centennial celebrations for Darwin's *Origin of Species* in 1959, the eminent historian of science C. C. Gillispie wrote: "Life is a purely physical phenomenon in Lamarck, and it is only because science has (quite rightly) left behind his conception of the physical that he has been systematically misunderstood and assimilated to a thesistic or vitalistic tradition which in fact he held in abhorrence."

Lamarck depicted his two sets of evolutionary forces as clearly distinct and destined to serve contrasting ends. The beauty of his theory—the embodiment of his *esprit de système*—lies in this clean contrast of both geometry and mechanism. The first set works upward to build progress in a strictly linear series of major anatomical designs (phyla) by recruiting mechanisms inherent in the structure and motion of living matter. The second set works sideward to extract branches made of individual lineages (species and genera) that respond to the influence of external circumstances by precise adaptations to particular environments. (These side branches may be visualized as projecting at right angles, perpendicular to the main trunk of progress. Vectors at right angles are termed orthogonal, and are mathematically independent, or uncorrelated).

Lamarck made this contrast explicit by stating that animals would form only a single line of progress if the pull of environmental adaptation did not interrupt, stymie, and divert the upward flow in particular circumstances:

> If the factor which works incessantly to complicate organization were the only one which had any influence on the shape and organs of animals, the growing complexity of organization would everywhere be very regular. But it is not; nature is forced to submit her works to the influence of their environment. . . . This is the special factor which occasionally produces . . . the often curious deviations that may be observed in the progression. (*Philosophie Zoologique,* 1809)

Thus, the complex order of life arises from the interplay of two forces in conflict, with progress driving lineages up the ladder and adaptation forcing them aside into channels set by the peculiarities of local environments:

> The state in which we find any animal is, on the one hand, the result of the increasing complexity of organization tending to form a regular gradation; and, on the other hand, of the influence of a multitude of very various conditions ever tending to destroy the regularity in the gradation of the increasing complexity of organization. (*Philosophie zoologique,* 1809)

Finally, in all his major evolutionary works, culminating in his multivolume treatise on invertebrate anatomy, Lamarck honored the first set of linear forces as primary and identified the second set as superposed and contrary—as in this famous statement, where he brands the lateral pull of adaptation as foreign, accidental, interfering, and anomalous:

> The plan followed by nature in producing animals clearly comprises a predominant prime cause. This endows animal life with the power to make organization gradually more complex. . . . Occasionally a foreign, accidental, and therefore variable cause has interfered with the execution of the plan, without, however, destroying it. This has created gaps in the series, in the form either of terminal branches that depart from the series in several points and alter its simplicity, or of anomalies observable in specific features of various organisms. (*Histoire naturelle des animaux sans vertèbres,* 1815)

III. The Values of Changing Theories

CHARLES DARWIN BEGAN the closing paragraph of his *Origin of Species* with a wonderful line that has served as the general title for these essays since I began the series in 1974: "There is grandeur in this view of life. . . ." No thinking or feeling person can deny either nature's grandeur or the depth and dignity of our discovery that a history of evolution binds all living creatures together. But in our world of diverse passions and psychologies, primary definitions (and visceral feelings) about grandeur differ widely among students of natural history. Darwin emphasized the bounteous diversity itself, in all its buzzing and blooming variety—for the finale of his closing paragraph contrasts the "dullness" of repetitive planetary cycling with the endless expansion and novelty of evolution's good work: "Whilst this planet has gone cycling on according to the fixed law of gravity, from so simple a beginning endless forms most beautiful and most wonderful have been, and are being, evolved."

But I suspect that Lamarck, following his own upbringing in the rigorous traditions of French rationalism during the Enlightenment, construed the definition of grandeur quite differently. As a devotee of *l'esprit de système*, Lamarck surely viewed the capacity of the human mind (his own in this case, for he was not a modest man) to apprehend the true and complete system of nature's rational order as the most remarkable feature of all. Thus, the logical clarity of the two-factor theory—with the primary cause establishing a linear march of rational progress and the opposed and subsidiary cause generating a more chaotic forest of adaptive diversity—must have struck Lamarck as the defining component both of nature's grandeur and of the power of evolution.

Our understanding of nature must always reflect a subtle interaction between messages from genuine phenomena truly out there in the real world, and the necessary filtering of such data through all the foibles and ordering devices internal to the human mind and its evolved modes of action. We cannot comprehend nature's complexity—particularly for such comprehensive subjects as evolution and the taxonomic structure of nature's diversity—unless we impose our mental theories of order upon the overt chaos that greets our senses. The different styles followed by scientists to balance and reconcile these two interacting (but partly contradictory) sources of order virtually define the rich variety of fruitful approaches pursued by a profession too often, and falsely, caricatured as a monolithic enterprise committed to a set of fixed procedures called *the* scientific method. Dangers and opportunities attend an overemphasis on either side. Rigid systematizers often misconstrue natural patterns by forcing their observations into rigidly preconceived structures of explanation. But colleagues of the opposite bent, who try to approach nature on her own terms, without preferred hypotheses to test, risk being either overwhelmed by a deluge of confusing information or falling prey to biases that become all the more controlling by their unconscious (and therefore unrecognized) status.

In this spectrum of useful approaches, Lamarck surely falls into the domain of scientists who place the logical beauty of a fully coherent theory above the messiness of nature's inevitable nuances and exceptions. In this context, I am all the more intrigued by Lamarck's later intellectual journey, so clearly contrary to his inclinations and inspired (in large part) by his inability to encompass new discoveries about the anatomy of invertebrates within the rigid confines of his beautiful system.

Nothing in the history of science can be more interesting or instructive than the intellectual drama of such a slow transformation in a fundamental view of life—from an initial recognition of trouble, to attempts at accommodation within a preferred system, then to varying degrees of openness toward substantial change, and sometimes even to full conversion. I particularly like to contemplate the contributions of external and internal factors to such a change: new data mounting a challenge from the outside, coordinated with an internal willingness to follow the logic of an old system to its points of failure and then to construct a revised theory imposing a different kind of consistency upon an altered world (with minimal changes if you remain in love with your previous certainties and tend to follow conservative intellectual strategies, or with potentially revolutionary impact if your temperament leads to iconoclasm and adventure). Reward and risk go hand in hand, for the great majority of thoroughly radical revisions must fail, even though the sweetest fruits await the few victors in this chanciest and most difficult of all mental battles.

When we can enjoy the privilege of watching a truly great intellect struggling with the most important of all biological concepts at a particularly interesting time in the history of science, then all factors coincide to generate a wonderful story offering unusual insight into the workings of science as well. When we can also experience the good fortune of locating a previously missing piece of history—in this case, the first record of a revision that would eventually alter the core of a central theory, although Lamarck, at this inception, surely had no inkling of how vigorously such a small seed could grow—we then gain the further blessing of an intriguing particular (the substrate of all good gossip) grafted onto an important generality. The prospect of being an unknown witness—the "fly on the wall" of a common metaphor—has always excited our fancy. And the opportunity to intrude upon a previously undocumented beginning—to be "present at the creation," in another common description—evokes an extra measure of charm. In this case, we begin with something almost inexpressibly humble—the classification of worms—and end with both a new geometry for animal life and a revised vision of evolution itself.

IV. Lamarck Emends His First Evolutionary Treatise

ONCE UPON A TIME, in a faraway world before the electronic revolution, and even before the invention of typewriters, authors submitted literal manuscripts (from the Latin for "written by hand") to their publishers. When writers revised a book for a second edition, they often worked from a specially prepared "interleaved copy," containing a blank sheet after each printed page. Corrections and additions could then be written on the blanks, enabling publishers to set a new edition from a coherent, bound document rather than from a confusing mess of loose or pasted insertions. Lamarck owned an interleaved copy of his first evolutionary treatise—the *Système des animaux sans vertèbres* of 1801. Although he never published a second edition, he did make comments on the blank pages—and he incorporated some of these statements into later works, particularly the *Philosophie zoologique* of 1809. This copy, which might have tempted me to a Faustian form of collusion with Mephistopheles if the opportunity had been offered, recently sold at auction for more money than even a tolerably solvent professor could ever dream of having at his disposal. But I was able to play the intellectual's usual role of voyeur during the few days of previewing before the sale, and I did recognize, in a crucial note in Lamarck's hand, a significance that had eluded previous observers. The eventual buyer (still unknown to me) expressed gratitude for the enhanced importance of his purchase, and he kindly offered (through the bookseller who had acted as his agent) to lend me the volume for a few days and allow me to publish the key note in this forum. I floated on cloud nine and happily rooted like a pig in . . . during those lovely days when I could hold and study my profession's closest approach to a holy grail.

Lamarck did not make copious additions on the interleaved copy, but several of his notes offer intriguing insights, while their general tenor teaches us something important about the relative weighting of his concerns. The first 48 pages of the printed book contain the Floréal address, Lamarck's initial statement of his evolutionary theory; the final 350 pages present a systematic classification of invertebrates, including a discussion of principles and a list and description of all recognized genera, treated phylum by phylum. Of Lamarck's thirty-seven handwritten additions on the blank pages, twenty-nine offer only a word or two and represent the ordinary activity of correcting small errors, inserting new information, or editing language. Lamarck makes fifteen comments about anatomy, mostly in his chapter on the genera of mollusks, the group he knew best. A further nine comments treat taxonomic issues of naming (adding a layman's moniker to the formal Latin designation, changing the name or affiliation of a genus); two add bibliographic data; and the final three edit some awkward language. Taken as an ensemble, I regard these comments as informative in correcting a false impression that Lamarck, by this time of seniority in his career, cared only for general theory and not for empirical detail. Clearly

Lamarck continued to cherish the minutiae of raw information and to keep up with developing knowledge—the primary signs of an active scientific life.

Of the eight longer comments, four appear as additions to the Floréal address. They provide instructive insight into Lamarck's character and concerns by fulfilling the conservative function of making more explicit—and elaborating by hypothetical examples—the central feature of his original evolutionary theory: the sharp distinction between causes of upward progress and sideward adaptation to local circumstances. In the two comments that attracted the most attention from potential buyers, Lamarck added examples of adaptation to local environments by inheritance of acquired characters: first, flatfishes living in shallow water that flatten their body to swim on their side and then move both eyes to the upper surface of their head; and second, snakes that move their eyes toward the top of their head since they live so close to the ground and must therefore perceive a world of danger above them, but that then need to develop a long and sensitive forked tongue to perceive the trouble in front that the eyes can no longer see. With these examples, Lamarck generalized his second set of forces by extending his stories to a variety of organisms; the Floréal address had confined itself to illustrations drawn from the habits and anatomy of birds. (The purely speculative character of these examples also helps us to understand why more sober empiricists, like Darwin and Cuvier, felt so uncomfortable with Lamarck's supposed data for evolution.) In any case, Lamarck published both examples almost verbatim in his *Philosophie zoologique* of 1809. A third comment then strengthens the other, and primary, linear force by arguing that the newly discovered platypus of Australia could link the penultimate group (birds) to the highest group (mammals). The fourth comment tries to explain the mechanisms of use and disuse by the differential flow of fluids through bodies.

The other set of four longer comments adorns the second part of the book, on taxonomic ordering of invertebrate animals. One insertion suggests that a small and enigmatic egg-shaped fossil should be classified within the phylum of corals and jellyfishes. A second statement, of particular interest to me, revises Lamarck's description of the clam genus *Trigonia*. This distinctive mollusk had long been known as a prominent fossil in Mesozoic rocks, but no Tertiary fossils or living specimens had ever been found, and naturalists therefore posited an extinction for *Trigonia* in the great Cretaceous mass dying that also wiped out the dinosaurs. But two French naturalists then found a living species of *Trigonia* in Australian waters in 1802, and Lamarck himself published the first description of this triumphant rediscovery in 1803. (As an undergraduate, I did my first technical research—and also wrote my first paper in the history of science—under the direction of Norman D. Newell at the American Museum of Natural History. He gave me a half-dozen, still preciously rare, specimens of modern Australian trigonians. When I gulped and admitted that I had no experience with dissection and feared butchering such a valuable bounty, he said to me, in his laconic manner—so inspirational for self-motivated students but so terrifying for the insecure—"Go down to the Fulton Fish Market and buy a bunch of quahogs. Practice on them first." I was far more terrified than inspired, but all's well that ends well.)

The final two comments provide the greatest visceral pleasure of all, because Lamarck added drawings to his words. The first sketch affirms Lamarck's continuing commitment to detail and to following and recording new discoveries. A small, white, delicate coiled shell of a cephalopod mollusk (the group including squid and octopuses) frequently washes up on beaches throughout the world. On the basis of this structure, Lamarck had proposed the genus *Spirula* in 1799, but the animal that makes the shell had never been found. As a particular mystery, no one knew whether the animal lived inside the shell (as in a modern chambered nautilus) or grew the shell within its body (as in the cuttlebone of a modern squid). The delicacy of the structure suggested a protected internal status, but the issue remained open. Soon after Lamarck's book appeared, naturalists discovered the animal of *Spirula* and affirmed an internal shell—a happy resolution that inspired Lamarck to a rare episode of artistic activity.

The last—and, as I here suggest, by far the most important—comment appears on the blank sheet following page 330, which contains descriptions of two remarkably different genera of worms—the medicinal leech *Hirudo* and the pond worm *Planaria*, known to nearly anyone who ever took a basic laboratory course in biology. Here Lamarck draws a simple sketch of the circulatory system of an annelid worm and then writes the following portentous words:

observation sur l'orgon des vers, dans les vers annelés et qui ont des organs extremes, le sang est rouge et circule dans des vaisseaux arteriels et veineux. leur organisation les place avant les insectes. les vers intestins doivent seuls se trouver après les insects. ils n'ont qu'un fluide blanc, libre, non contenu dans des vaisseaux. Cuvier. extrait d'un mém. lu a l'institut le 11 nivose an 10. (Observation on the organization of worms. In annelid worms, which have external organs, the blood is red and circulates in arterial and venous vessels. Their organization places them before the insects. Only the internal worms come after the insects. They have only a white fluid, free, and not contained in vessels. Cuvier. Extract from a memoir read at the Institute on the 11th day of the month of Nivôse year 10 of the Republic.)

Clearly Lamarck now recognizes a vital distinction between two groups that had once been lumped together into the general category of worms. He regards one group—the annelids, including earthworms, leeches, and the marine polychaetes—as highly advanced, even more so than insects. (Lamarck usually presented his scale of animal life from the top down, starting with humans and ending with infusorians, rather than following the later convention of working upward. Thus, he states that annelids come before insects because they are more advanced—that is, closer to the mammalian top.) But another group, the internal worms* (mostly parasites living within the bodies of other animals), rank far lower on the scale—even after (that is, anatomically simpler than) insects. These two distinct groups, previously conflated, must now be widely separated in the taxonomic ordering of life. Ironically, Lamarck acknowledges his colleague Cuvier (who would later turn against him and virtually destroy his reputation) as the source for a key item of information that changed his mind—Cuvier's report (presented at a meeting during the winter of 1801–1802, soon after the publication of Lamarck's book) that annelids possessed a complicated circulatory system, with red blood running in arteries and veins, whereas internal worms grew no discrete blood vessels and moved only a white fluid through their body cavity.

Obviously Lamarck viewed this new information as especially important, for no other anatomical note receives nearly such prominence in his additions, while only one other observation (a simple new bit of information, without much theoretical meaning) merits a drawing. But why did Lamarck view this division of worms as so important? And how could such an apparently dull and technical decision about naming act as a pivot and initiator for a new view of life?

V. An Odyssey of Worms

I HAVE ALWAYS considered it odd (and redolent either of arrogance or parochialism) when a small minority divides the world into the two wildly unbalanced categories of itself versus all others and then defines the big category

---

*The standard English literature on this subject always translates Lamarck's phrase incorrectly as "intestinal worms." These parasites dwell in several organs and other locations in vertebrate (and other) bodies, not only in the intestines. In French, the word *intestin* conveys the more general meaning of "internal" or "inside."

as an absence of the small—as in my grandmother's taxonomy for *Homo sapiens:* Jews and non-Jews. Yet our conventional classification of animals follows the same strategy by drawing a basic distinction between vertebrates and invertebrates—when only about forty thousand of more than a million named species belong to the relatively small lineage of vertebrates.

On the venerable principle that bad situations can always be worse, we can gain some solace by noting the even greater imbalance devised by the founder of modern taxonomy, Carolus Linnaeus. At least we now recognize vertebrates as only part of a single phylum, while most modern schemes divide invertebrates into some twenty to thirty separate phyla. But in his *Systema naturae* of 1758, the founding document of modern zoological nomenclature, Linnaeus identified only six basic animal groups: four among vertebrates (mammals, birds, reptiles, and fishes) and two for the entire realm of invertebrates (Insecta, for insects and their relatives, and Vermes, literally "worms," for nearly everything else).

When Lamarck became professor of invertebrates at the Muséum d'Histoire Naturelle in 1793 (with an official title in a Linnaean straitjacket as professor of insects, and worms), he already recognized that reform demanded the dismemberment of Linnaeus's "wastebucket" category of Vermes. (*Wastebucket* actually ranks as a semitechnical term among professional taxonomists, a description for inflated groups that become receptacles for heterogeneous bits and pieces that most folks would rather ignore—as in Linnaeus's relegation of all "primitive" animals to the category of "worms," ranking far beneath the notice of specialists on vertebrates.)

In his 1801 book, *Système des animaux sans vertèbres,* Lamarck identified the hodgepodge of Linnaeus's Vermes as the biggest headache and impediment in zoology:

> The celebrated Linnaeus, and almost all other naturalists up to now, have divided the entire series of invertebrate animals into only two classes: insects and worms. As a consequence, anything that could not be called an insect must belong, without exception, to the class of worms.

By the time Lamarck wrote his most famous book, *Philosophie zoologique,* in 1809, his frustration had only increased; he called Linnaeus's class of worms "une espèce de chaos dans lequel les objets très disparates se trouvent réunis (a kind of chaos where very disparate objects have been united together)." He then blamed the great man himself for this sorry situation: "The authority of this scientist carried such great weight among naturalists that no one dared to change this monstrous class of worms." (By writing *cette classe monstrueuse*, I am confident that Lamarck meant to attack the sheer numbers of included genera, not the moral status, of Linnaeus's Vermes). Lamarck therefore began his campaign of reform by raiding Vermes and gradually adding the extracted groups as novel phyla to his newly named category of invertebrates. In his first museum lecture course in 1793, he had already expanded the Linnaean duality to a ladder of progress with five rungs—mollusks, insects, worms, echinoderms, and polyps (corals and jellyfish)—by liberating three new phyla from the wastebucket of Vermes.

This reform accelerated in 1795, when Georges Cuvier arrived at the museum and began to study invertebrates as well. The two men collaborated in friendship at first, and they surely operated as one mind on the key issue of dismembering Vermes. Thus, during almost every annual course of lectures, Lamarck continued to add phyla, extracting most new groups from Vermes but some from the overblown Linnaean Insecta as well. In year 7 of the French revolutionary calendar (1799), he established the Crustacea (for marine arthropods, including crabs, shrimps, and lobsters), and in year 8 (1800), the Arachnida (for spiders and scorpions). Lamarck's invertebrate classification of 1801 therefore featured a growing ladder of progress, now bearing seven rungs. In 1809 he presented a purely linear sequence of progress for the last time in his most famous book, *Philosophie zoologique*. His tall and rigid ladder now contained fourteen rungs, as he had added the four traditional groups of vertebrates atop a list of invertebrate phyla that had just reached double digits.

So far, Lamarck had done nothing to inspire any reconsideration of the evolutionary views first presented in his Floréal address of 1800. His taxonomic reforms, in this sense, had been entirely conventional in adding weight and strength to his original views. The Floréal statement had contrasted a linear force leading to progress in major groups with a lateral force causing local adaptation in particular lineages. In 1800 his ladder had included only seven groups. By 1809 he had doubled its length while preserving the same strictly linear form—thus strengthening his central contrast between two forces by granting the linear impetus a greatly expanded field for its inexorably exceptionless operation.

But if Lamarck's first reform of Linnaeus—the expansion of groups into a longer linear series—had conserved and strengthened his original concept of evolution, he now embarked upon a second reform, destined (though he surely had no inkling at the outset) to yield the opposite effect of forcing a fundamental change in his view of life. He had, heretofore, only extracted misaligned groups from Linnaeus's original Vermes. He now needed to consider the core of Vermes itself and to determine whether waste and rot existed at the foundation as well.

"Worms," in our vernacular understanding, are defined both broadly and negatively (bad criteria spelling inevitable trouble down the road) as soft-bodied, bilaterally symmetrical animals, roughly cylindrical in shape and lacking appendages or prominent sense organs. By these criteria, both earthworms and tapeworms fill the bill. For nearly ten years, Lamarck did not seriously challenge this core. But he could not permanently ignore the glaring problem (recognized but usually swept under the rug by naturalists) that this broad vernacular category seemed to include at least two kinds of organisms bearing little relationship beyond a superficial and overt similarity of external form. On the one hand, a prominent group of free-living creatures—earthworms and their allies—built bodies composed of rings or segments and also developed internal organs of substantial complexity, including nerve tubes, blood vessels, and a digestive tract. But another assemblage of largely parasitic creatures—tapeworms and their allies—grew virtually no discretely recognizable internal organs at all and therefore seemed much "lower" than earthworms and their kin by Lamarck's own favored concept of an organic scale of complexity. Would the heart of Vermes therefore need to be dismembered as well?

This problem was already worrying Lamarck when he published the Floréal address in his 1801 compendium on invertebrate anatomy, but he was not yet ready to impose a formal divorce upon the two basic groups of "worms." Either standard of definition, taken by itself—different anatomies *or* disparate environments—might not offer sufficient impetus for taxonomic separation. But the two criteria correlated perfectly in the remaining Vermes: the earthworm group possessed complex anatomy *and* lived freely in the outside world; the tapeworm group maintained maximal simplicity among mobile animals *and* lived almost exclusively within the bodies of other creatures. Lamarck therefore opted for an intermediary solution. He would not yet dismember Vermes but would establish two subdivisions within the class: *vers externes* (external worms) for earthworms and their allies and *vers intestins* (internal worms) for tapeworms and their relatives. He stressed the simple anatomy of the parasitic subgroup and defended their new name as a spur to further study, while arguing that knowledge remained insufficient to advocate a deeper separation:

> It is very important to know them [the internal worms], and this name will facilitate their study. But aside from this motive, I also believe that such a di-

vision is the most natural . . . because the internal worms are much more im-
perfect and simply organized than the other worms. Nevertheless, we know so
little about their origin that we cannot yet make them a separate order.

At this point the crucial incident occurred that sparked Lamarck to an ir-
revocable and cascading reassessment of his evolutionary views. He attended
Cuvier's lecture during the winter of 1801–1802 (year 10 of the revolutionary
calendar) and became convinced by his colleague's elegant data on the
anatomy of external worms that the differences between his two subdivisions
were too great to permit their continued residence in the same class. He would,
after all, have to split the heart of Vermes. In his next lecture course, in the
spring of 1802, Lamarck formally established the class Annelida for the external worms (retaining Vermes for the internal worms alone) and then separated
the two classes widely by placing his new annelids above insects in linear complexity while leaving the internal worms near the bottom of the ladder, well
below insects.

Lamarck formally acknowledged Cuvier's spur when he wrote a history
of his successive changes in classifying invertebrates for the *Philosophie zoologique* of 1809:

> Mr. Cuvier discovered the existence of arterial and venous vessels in distinct
> animals that had been confounded with other very differently organized an-
> imals under the name of worms. I soon used this new fact to perfect my clas-
> sification; therefore, I established the class of annelids in my course for year
> 10 [1802].

The handwritten note and drawing in the interleaved copy of Lamarck's
1801 book tells much the same story—but what a contrast, in both intellectual
and emotional intrigue, between a sober memory written long after an inspi-
ration and the inky evidence of the moment of enlightenment itself!

But this tale should now be raising questions in the minds of readers. Why
am I making such a fuss about this particular taxonomic change: the final division of Vermes into a highly ranked group of annelids and a primitive class
of internal worms? In what way does this alteration differ from any other
change in classification previously discussed? In all cases, Lamarck subdivided
Linnaeus's class Vermes and established new phyla in his favored linear series,
thus reinforcing his view of evolution as built by contrasting forces of linear
progress and lateral adaptation. Wasn't he just following the same procedure
in extracting annelids and placing them on a new rung of his ladder? So it

might seem—at first. But Lamarck was too smart, and too honorable, to ignore a logical problem directly and inevitably instigated by this particular division of worms—and the proper solution broke his system.

At first, Lamarck did treat the extraction of annelids as just another addition to his constantly improving linear series. But as the years passed, he became more and more bothered by an acute problem evoked by an inherent conflict between this taxonomic decision and the precise logic of his overarching system. Lamarck had ranked the phylum Vermes, now restricted to the internal worms alone, just above a group he named *radiaires* (radially symmetrical animals)—actually (by modern understanding) a false amalgam of jellyfishes from the coelenterate phylum and sea urchins and their relatives from the echinoderm phylum. Worms had to rank above radiates because bilateral symmetry and directional motion trump radial symmetry and an attached (or not very mobile) lifestyle—at least in conventional views about ladders of progress (which, of course, use mobile and bilaterally symmetrical humans as the ultimate standard). But the parasitic internal worms lack the two most important organ systems—nerve ganglia and cords, and circulatory vessels—that virtually define complexity on the traditional ladder. Yet echinoderms, within the "lower" radiate phylum, develop both nervous and circulatory systems. They circulate seawater rather than blood, but they do run their fluids through tubes.

If Lamarck's primary "force that tends incessantly to complicate organization" truly works in a universal and exceptionless manner, then how can such an inconsistent situation arise? If the force be general, then any given group must stand fully higher or lower than any other. A group cannot be higher for some features but lower for others. Taxonomic experts cannot pick and choose. He who lives by the line must die by the line.

This problem did not arise so long as annelids remained in the class of worms. Lamarck, after all, had never argued that each genus of a higher group must rank above all members of a lower group in every body part. He only claimed that the "principal masses" of organic design run in pure linear order. Individual genera may degenerate or adapt to less complex environments—but so long as some genera display the higher conformation in all features, then the entire group retains its status. In this case, so long as annelids remained, then many worms possessed organ systems more complex than any comparable structure in any lower group—and the entire class of worms could retain its unambiguous position above radiates and other primitive forms. But with the division of worms and the banishment of complex annelids, Lamarck now faced the logical dilemma of a coherent group (the internal parasitic worms) standing higher than radiates in some key features but lower in others. The pure march of nature's progress—the keystone of Lamarck's entire system—had been fractured.

Lamarck struggled with this problem for several years. He stuck to the line of progress in 1802 and again—for the last time, and in a particularly uncompromising manner that must, in retrospect, have been a last hurrah before the fall—in the first volume of his seminal work, the *Philosophie zoologique* of 1809. But honesty eventually trumped hope. Just before publication, Lamarck appended a short chapter of additions to volume two of the *Philosophie zoologique*. He now, if only tentatively, floated a new scheme that would resolve his problem with worms but also unravel his precious linear system.

Lamarck had always argued that life began with the spontaneous generation of infusorians (single-celled animals) in ponds. But suppose that spontaneous generation occurs twice and in two distinct environments—in the external world, for a lineage beginning with infusorians, and inside the bodies of other creatures, for a second lineage beginning with worms? Lamarck therefore wrote that "worms seem to form one initial lineage in the scale of animals, just as, evidently, the infusorians form the other branch."

Lamarck then faced the problem of allocating the higher groups. To which of the two great lines does each belong? He presented his preliminary thoughts in a chart—perhaps the first evolutionary branching diagram ever published in the history of biology—that directly contradicted his previous image of a single ladder. Lamarck begins (at the top, contrary to current convention) with two lines, one labeled *infusoires* (single-celled animals) and the other *vers* (worms). He then inserts light dots to suggest possible allocations of the higher phyla to the two lines. The logical problem that broke his system has now been solved—for the *radiaires*, standing below worms in some features but above in others, now rank in an entirely separate series, directly following an infusorian beginning.

When mental floodgates open, the tide of reform must sweep to other locales. Once he had admitted branching and separation at all, Lamarck could hardly avoid the temptation to apply this new scheme to other old problems. Therefore, he also suggested some substantial branching at the end of his array. He had always been bothered by the conventional summit of reptiles to birds to mammals, for birds seemed just different from, rather than inferior to, mammals. Lamarck therefore proposed (and drew on his revolutionary chart) that reptiles branched at the end of the series, one line passing from turtles to birds *(oiseaux)* to *monotrèmes* (platypuses, which Lamarck now considered as separate from mammals), and the other from crocodiles to marine mammals (labeled *m. amphibies*) to terrestrial mammals. Finally, and still in the new spirit, he even posited a threefold branching in the transition to terrestrial mammals, leading to separate lines for whales (*m. cétacés*), hoofed animals (*m. ongulés*), and mammals with nails (*m. onguiculés*), including carnivores, rodents, and primates.

Finally, Lamarck explicitly connected the two reforms: the admission of two sequences of spontaneous generation at the bottom and a branching among higher vertebrates at the top: "The animal scale begins with at least two branches; in the course of its extent, several branches seem to end in different places."

After the *Philosophie zoologique* of 1809, Lamarck wrote one additional major book on evolution, the introductory volume (1815) of his *Histoire naturelle des animaux sans vertèbres*. Here he transcended all the tentativeness of his 1809 revision and announced his conversion to branching as the fundamental pattern of evolution. In direct contradiction to the linear model that had shaped all his previous work, Lamarck stated simply and without ambiguity: "Dans sa production des differents animaux, la nature n'a pas executé une série unique et simple. (In its production of the different animals, nature has not fashioned a single and simple series.)"

He then emphasized the branching form of his new model and explained how the division of worms, inspired by Cuvier's observations, had broken his former system and impelled his revision:

> The order is far from being simple; it is branching (*rameux*) and even appears to be constructed of several distinct series. . . . The animals that belonged to the class of worms display a great disparity of organization. . . . The most imperfect of these animals arise by spontaneous generation, and the worms [now restricted to *vers intestins*, with annelids removed] truly form their own series, later in origin than the one that began with infusorians.

Lamarck's third and last chart shows how far he had progressed both in his own confidence and in copious branching on his new tree of life. He titles the chart "Presumed order of the formation of animals, showing two separate and sub-branching series." Note how the two major lines of separate spontaneous generation—one beginning with infusorians, the other with internal worms—are now clearly marked and separated. Note also how each series also divides within itself, thus establishing the process of branching as a key theme at all scales of the system. The infusorian line branches at the level of polyps (corals and jellyfish) into a line of radiates and a line terminating in mollusks. The second line of worms also branches in two, leading to annelids on one side and insects on the other. But the insect line then splits again (a tertiary division) into a lineage of crustaceans and barnacles (labeled *cirrhipèdes*) and another of arachnids (spiders and scorpions).

Finally, we must recognize that these major changes do not affect only the overt geometry of animal organization. The conversion from linearity to branching also—and perhaps even more importantly—marks a profound shift in Lamarck's underlying theory of nature. He had based his original system, defending it vociferously until 1809, on a fundamental division of two independent forces—a primary cause that builds basic anatomies in an unbroken line of progress and a subsidiary lateral force that draws single lineages off the line into byways of immediate adaptation to local environments. A set of philosophical consequences then spring from this model: the predictable and lawlike character of evolution lies patent in the primary force and its ladder of progress reaching to man, while accidents of history (leading to local adaptations) can then be dismissed as secondary and truly independent from the overarching order.

But the branching system destroys this neat and comforting scheme for two major reasons. First of all, the two forces become intermingled and conflated in the branching itself. We can no longer distinguish two independent and orthogonal powers working at right angles. Progress may occur along any branch, to be sure, but the very act of division implies an environmental impetus to split the main line—and Lamarck had always advocated a complete and principled distinction between a single and inexorable main line and the numerous minor deviations that can draw off a long-necked giraffe or an eyeless mole but can never disrupt or ramify the major designs of animal life. In the new model, however, environment intrudes at the first construction of basic order—as one group arises spontaneously in ponds, and another inside the bodies of other creatures! Moreover, each of the two resulting lines then branches further, and unpredictably, under environmental impetuses that were not supposed to derail the force of progress among major groups, as when insects split into a terrestrial line of arachnids and a marine line leading to crustaceans and barnacles.

Secondly, the forces of history and natural complexity had now triumphed over the scientific ideal of a predictable and lawlike system. The taxonomy of animals could no longer embody an overarching plan of progress, illustrating the fundamental order, harmony, and predictable good sense of the natural world (and perhaps even the explicit care of a loving deity, whose plans we may hope to understand because he thinks as we do). Now the confusing, particular, local, and unpredictable forces of complex environments held sway, ready at any time to impose a deviation upon any group with enough hubris to suppose that Emerson's forthcoming words could describe their inevitable progress:

> And, striving to be man,
>   the worm
> Mounts through all the
>   spires of form.

VI. Lamarck's Epilogue and My Own

FOLLOWING HIS LAST and greatest treatise on the anatomy of invertebrate organisms, Lamarck published only one other major work: *Analytic System of Positive Knowledge of Man* (1820). This rare book has not been consulted by previous historians who traced the development of Lamarck's changing views on the classification of animals. Thus, traditional accounts stop at Lamarck's 1815 revision, with its fundamental distinction between two separate lineages of spontaneous generation. The impression therefore persists that Lamarck never fully embraced the branching model, later exemplified by Darwin as the "tree of life," with a common trunk of origin for all creatures and no main line of growth thereafter. Lamarck had compromised his original ladder of progress by advocating two separate origins for living things, but he could continue to stress linearity in each of the resulting series.

But his 1820 book, although primarily a treatise on psychology, does include a chapter on the classification of animals, and I was excited to find that Lamarck does pursue his revisionary path further and does finally arrive at a truly branching model for a tree of life. Moreover, in a remarkable passage, Lamarck also recognizes the philosophical implications of his full switch by acknowledging a reversal in his ranking of natural forces—one of the most interesting (and honorable) intellectual conversions I have ever read.

Lamarck still talks about forces of progress and forces of branching, and he does argue that progress will proceed along each branch. But branching has triumphed as a primary and controlling theme, and Lamarck now frames his entire discussion of animal taxonomy in terms of successive points of division. For example, consider this epitome of vertebrate evolution:

> Reptiles come necessarily after fishes. They build a branching sequence, with one branch leading from turtles to platypuses to the diverse group of birds,

while the other seems to direct itself, via lizards, toward the mammals. The birds then . . . build a richly varied branching series, with one branch ending in birds of prey.

In previous models, he had viewed birds of prey as the top rung of a single avian ladder.

But much more radically, his 1815 model, based on two lines of spontaneous generation, has now disappeared. In its place, Lamarck advocates the same tree of life that would later become conventional through the influence of Darwin and other early evolutionists. Lamarck now proposes a single common ancestor for all animals, called a monad. From this beginning, infusorians evolve, followed by polyps, arising "directly and almost without a gap." But polyps then branch to build the rest of life's tree. "Instead of continuing as a single series, the polyps appear to divide themselves into three branches": the radiates, which end without evolving any further; the worms, which continue to branch into all phyla of segmented animals, including annelids, insects, arachnids, crustaceans, and barnacles, each by a separate event of division; and the tunicates (now regarded as marine organisms closely related to vertebrates), which later split to form several lines of mollusks and vertebrates.

Lamarck then acknowledges the profound philosophical revision implied by a branching model for nature's fundamental order. He had always viewed the linear force of progress as primary. As late as 1815, even after changing his model to permit extensive branching and two environmentally induced sequences of spontaneous generation, Lamarck continues to emphasize the primary power of the linear force compared with disturbing and anomalous exceptions produced by lateral environmental causes, called *l'influence des circonstances*. I quoted his key passage in Section II of this essay:

> The plan followed by nature in producing animals clearly comprises a predominant prime cause. This endows animal life with the power to make organization gradually more complex. . . . Occasionally a foreign, accidental, and therefore variable cause has interfered with the execution of the plan . . . [producing] branches that depart from the series in several points and alter its simplicity.

But five years later, in his final book of 1820, Lamarck now rejects this controlling concept of his career and embraces the opposite conclusion. The influence of circumstances (leading to a branching model of animal taxonomy) rules the paths of evolution. All general laws—of progress or anything else—

must be subservient to the immediate singularities of environments and histories. The influence of circumstances has risen from the position of a disturbing and peripheral joker to true lord-of-all (with an empire to boot):

> Let us consider the most influential cause for everything done by nature, the only cause that can lead to an understanding of everything that nature produces. . . . This is, in effect, a cause whose power is absolute, superior even to nature, since it regulates all nature's acts, a cause whose empire embraces all parts of nature's domain. . . . This cause resides in the power that circumstances have to modify all operations of nature, to force nature to change continually the laws that she would have followed without [the intervention of] these circumstances, and to determine the character of each of her products. The extreme diversity of nature's productions must also be attributed to this cause.

Lamarck's great intellectual journey began with a first public address about evolution, delivered in 1800 during a month that the revolutionary government had auspiciously named Floréal, or flowering. He then developed the first comprehensive theory of evolution in modern science—an achievement that won him a secure place in any scientific hall of fame or list of immortals—despite the vicissitudes of his reputation during his own lifetime and immediately thereafter.

But Lamarck's original system failed—and not for the reasons we specify today in false hindsight (the triumph of Mendelism over Lamarck's false belief in inheritance of acquired characters) but from inconsistencies that new information imposed upon the central logic of Lamarck's system during his own lifetime. We can identify a fulcrum, a key moment, in the unraveling of Lamarck's original theory: when he attended a lecture by Cuvier on the anatomy of annelids and recognized that he would have to split his taxonomic class of worms into two distinct groups. This recognition—which Lamarck recorded with excitement (and original art) as a handwritten insertion into his first published book on evolution—unleashed a growing cascade of consequences that, by Lamarck's last book of 1820, destroyed his original theory of primary ladders of progress versus subsidiary lateral deviations and led him to embrace the opposite model (in both geometry of animal classification and basic philosophy of nature) of a branching tree of life.

A conventional interpretation would view this tale as fundamentally sad, if not tragic, and would surely note a remarkable symbol and irony for a literary conclusion. Lamarck began his adventure in the springtime month of flowering. But he heard Cuvier's lecture, and his system began to crumble, on the eleventh day of Nivôse, the winter month of snow. How fitting—to begin with springtime joy and promise, and to end in the cold and darkness of winter. How fitting, in one distorted sense—but how very, very wrong. I do not deny or belittle Lamarck's personal distress, but how can we view his slow acknowledgment of logical error, and his willingness to construct an entirely new and contrary explanation, as anything other than a heroic act, worthy of our greatest admiration and identifying Lamarck as one of the finest intellects in the history of biology (a name that he invented for his profession)?

Two major reasons lead me to view Lamarck's intellectual odyssey in this eminently positive light. First, what can be more salutary in science than the flexibility that allows a person to change his mind, and to do so not for a minor point, under the compulsion of irrefutable data, but to rethink and reverse the most fundamental concept underlying a basic philosophy of nature. I would argue, secondly, that Lamarck's journey teaches us something truly important about the interaction between nature and our attempts to understand her ways. The fallacies and foibles of human thinking lead us into systematic and predictable trouble when we try to grasp the complexities of external reality. Among these foibles, our persistent attempts to build abstractly beautiful, logically impeccable, and comprehensively simplified systems always lead us astray. Lamarck far exceeded most of us in his attraction to this perilous style of theorizing—this *esprit de système*—and he therefore fell further and harder because he also possessed the honesty and intellectual power to probe his mistakes.

Nature, to cite a modern cliché, always bats last. She will not succumb to the simplicities of our hopes or mental foibles, but she remains eminently comprehensible. Evolution follows the syncopated drumbeats of complex and contingent histories, shaped by the vagaries and uniquenesses of time, place, and environment. Simple laws with predictable outcomes cannot fully describe the pageant and pathways of life. A linear march of progress must fail as a model for evolution, but a luxuriantly branching tree does capture the basic geometry of history.

When Lamarck snatched victory from the jaws of his defeat (by abandoning his beloved ladder of life and embracing the tree), he stood in proper humility before nature's complexity—a lesson for us all. But he also continued to wrestle with nature, to struggle to understand and even to tame her ways, not simply to bow down and acknowledge sovereignty. Only the most heroic people can follow Job's great example in owning error while continuing to hurl defiance and shout, "I am here." Lamarck greeted nature (traditionally construed as female) with Job's ultimate challenge to God (construed as male, in equally dubious tradition): "Though he slay me, yet will I trust in him; but I will maintain mine own ways before him" (Job 13:15).

I therefore propose that we reinterpret the symbolic meaning of Lamarck's undoing in the month of Nivôse. Cuvier's challenge elicited a cascade of discovery and reform, not the battering of bitter defeat. And snow also suggests metaphors of softness, whiteness, and purification—not only of frost, darkness, and destruction. God, in a much kinder mood than he showed to poor Job, promises his people in the first chapter of Isaiah: "Though your sins be as scarlet, they shall be as white as snow; though they be red like crimson, they shall be as wool" (Is. 1:18). We should also remember that this verse begins with an even more famous statement—a watchword for an intellectual life, and a testimony to Lamarck's brilliance and flexibility: "Come now, and let us reason together."

SUSAN MCCARTHY

Must Dog Eat Dog?

FROM *SALON*

The conventional narratives of evolution, whether put forward by scientists or popularizers, suggest that all living things are ruthlessly competitive, and that success comes only to the selfish. Reviewing recent research and literature on the subject, Susan McCarthy wittily asks whether these assumptions tell the whole evolution story.

Face facts: Men are nothing but horn-dogs, and women only want men for their money. Equations prove: No one ever does anyone a favor unless there's something in it for them. Accept the data: People are just bloodthirsty apes with a flair for spin. We're deluded puppets of our genes, and our genes like us deluded. We have met the enemy and it is our DNA.

Evolutionary psychologists revel in telling Uncomfortable Truths. So eager are many thinkers to face up to hard facts, that they may even face up to hard, fact-like objects that turn out not to *be* facts at all. But hey! We faced them bravely, and you should too.

Robert Ardrey and Desmond Morris, in books like *The Territorial Imperative* (1966) and *The Naked Ape* (1967) prepared the ground with their visions of killer apes with violent, lustful origins. But evolutionary psychology (an offshoot of sociobiology) really got rolling in the mid-1970s with E. O. Wilson's *Sociobiology* and Richard Dawkins' *The Selfish Gene.* In no time works of pop sociobiology made these already accessible works even more accessible. Theorists divided into hostile camps—like killer apes, but not so naked. Attackers and supporters of Wilson, of Dawkins, of Richard Lewontin, of Stephen Jay Gould formed up in short order. Wilson announced that the social sciences would have to knuckle under and become fiefdoms of imperial sociobiology. Some social scientists, especially psychologists, leapt at the idea of becoming Real Scientists; others drew back in outrage. Since then, the controversies of evolutionary psychology have raged in both the journals and the popular press.

People seldom enjoy thinking of their own motivations as anything but impeccably reasoned and noble, so there are plenty of uncomfortable truths for all of us to face. One of them is that we are animals, which is awkward, considering the snide things we've been saying about the other animals all this while.

But one of the uncomfortable truths that evolutionary psychologists must face is that we are also humans, deeply enmeshed in cultures of unprecedented complexity. Not only has culture created science, but the worldviews of scientists as well. It will always be hard to map the intertidal zone where our cultural attitudes end and our genetic heritage takes over, but recognizing this, and distinguishing between waterlines at high and low tide is an intellectual effort worth making.

One of evolutionary psychology's weaknesses is the lack of human data. Unable to experiment on this intriguing species, or observe them in culturally uncontaminated environments, scientists fall back on population analysis, surveys of dubious significance, and, all too often, common sense intuition about The Way Folks Are. The initial common sense intuition of the 1960s and 1970s seems to have combined the gender sensibilities of 1950s clip art with the paranoid Machiavellian worldview of an assistant professor unfairly denied tenure.

Interestingly, much of the criticism of evolutionary psychology's original vision has come from primatologists like Frans de Waal, Sarah Blaffer Hrdy, Patricia Gowaty, Barbara Smuts and Meredith Small, who have access to data about how some primate populations really work. The widespread assumption that dominant males sire all the infants in groups of social primates, for example, has proven remarkably shaky. Recent studies of primates from rhesus monkeys to chimpanzees have found females lusting after nerdy low-status monkeys, sneaking off with low-status guys, and even refusing to have anything to do with sand-kicking bullies.

Challenges to evolutionary psychology's earlier thinking about gender were inevitable. A glance through *The Moral Animal,* by Robert Wright, an award-winning science writer, not only displays widely annoying assumptions and pronouncements about how women are, but a stream of complaints about feminism (and vague, dishonest, disingenuous, intimidating feminists). Wright's gripes about feminism are only slightly more explicit than those of many theorists. He tops this with the ponderously playful suggestion that real feminism should favor polygamy because if some women were second, third (and so on) wives of tycoons, the remaining women would have more men to choose among, and there would be no single mothers. Curiously, America's women do not seem to have uttered a collective sigh of relief at having real feminism explained.

This socio-financial analysis is part of the gold-digger theory of femininity. It starts reasonably enough by pointing out that females put a larger physical investment into giving birth than males do. (In fact, the definition of female-ness isn't two X chromosomes: Biologists decide whether a bird or fish is female by determining which one produces the egg.) Sociobiological theory then ex-plores the implications of that differing investment. At some point, however, theory often jumps tracks, and heads into downtown with whistles blowing. "You bitch!" screams the engineer. "All you ever wanted was my money!"

The notion is that the ideal strategy for men is to have sex with as many women as possible, in order to get their genes into the next generation, while the ideal strategy for women is to commandeer a man's wealth to provide the resources to keep the kid alive, and this idea has been widely ballyhooed. In this view, monogamy is a perpetual tussle between these competing interests—if only we would admit it to ourselves.

Several years ago, an article by Lance Morrow in *Time* asserted that insights from animal behavior show that "females organize their lives around the get-ting of resources (food, shelter, nice things) while males organize themselves around the getting of females." Oddly, it does not appear that Morrow himself roams the streets, hungry and homeless, looking for women.

In *Woman: An Intimate Geography,* by Natalie Angier, several chapters grapple with received wisdoms of evolutionary psychology. Angier casts sub-stantial doubt on the idea that women evolved to nab rich guys. She argues that in the hunter-gatherer societies in which our tastes are presumed to have evolved, there often are no rich guys. Everyone is equally poor. Nor do women rely on men to provision them—everybody works. As for our earlier ape an-cestors, again females were probably self-supporting. The idea that nobody eats until he brings home the bacon is a new one, evolutionarily speaking.

In the hunt for data to support the idea that men only care about sex and women couldn't care less, there have been many idiot surveys, including the classic, endlessly reported one in which two charming experimenters, male and female, approached people of the opposite sex on a college campus and asked them to have sex. None of the women agreed to have sex with the male experimenter. Three quarters of the men said they would have sex with the female experimenter. Therefore, men are naturally sexually indiscriminate and women are naturally "coy." We have our proof!

The experimenters appear to have forgotten that people don't always tell inquisitive strangers what they're thinking, as pollsters have found to their regret. Experimenters can't know if people would really have had sex with them until they actually put down the clipboard and try, and that's apparently a test few researchers follow through on. Or if they do, they're not telling.

Angier also points out that random strangers requesting sex are apt to seem, and be, more physically dangerous to women than to men. So that even if a lady can think of no better study break than anonymous sex with Clipboard Boy, she may regretfully decline.

As more and more researchers are pointing out flaws in the views of gender originally presented by evolutionary psychology, it's annoying to note that the revisionism is mostly coming from women. This is supposed to be science, tested with actual data. Why do the women have to be the ones to point this stuff out? Significantly, a lot depends on where you stand, and sex roles appear more inevitable to the male evolutionary psychologists who first propounded their genetic immutability.

The other big trend in evolutionary psychology, which also comes from studying non-human primates, is about peacemaking and cooperation. Aggression, while real, isn't the entire story. Nature is red in tooth and claw, but it turns out there are also plenty of hugs to go around.

When Frans de Waal first noticed how hard chimpanzees and other (captive) primates work to keep things from getting out of hand in their groups, he was astonished to find that he was in unknown territory. Much had been written about violence and aggression of all kinds, but very little about how social animals go about the vital task of getting along.

De Waal, the author of *Peacemaking Among Primates,* initially had difficulty getting other primatologists to pay attention. They didn't even like the words. "Reconciliation" seemed to de Waal like a good word for the friendly overtures made to each other by apes who had been fighting minutes before, often promoted by a third ape with an interest in group harmony. "Couldn't you call it 'first post-aggression contact'?" colleagues pleaded. Similarly, de Waal notes, Barbara Smuts was reproached for writing of "friendship" in baboons.

De Waal's latest book, *Good Natured: The Origins of Right and Wrong in Humans and Other Animals,* describes some interesting research into the mutability of monkey nature. He tells of an experiment with rhesus monkeys and stump-tailed monkeys, species with different calls, gestures and styles. Stump-tails are more peaceable than rhesus, less obsessed with hierarchy and who sits where, and, having apparently read de Waal, they are very into reconciliation. After fights they reconcile three times as often as rhesus, and they have a big repertoire of reassuring gestures.

"Why can't a rhesus be more like a stump-tail?" was de Waal's question, and he set up a mixed colony. The stump-tails he chose were six months older than the rhesus, and a little bigger. At first the rhesus tried threatening the stump-tails, but they didn't respond. The two monkey species played separately and slept separately, in two big huddles. They groomed each other, however, a process encouraged by the fact that the stump-tails were enchanted by the opportunity to examine long rhesus tails.

Soon the rhesus and the stump-tails were playing and sleeping together in one monkey heap. Although the rhesus did not adopt stump-tail calls and gestures, they did succumb to their touchy-feely culture and increase their rate of reconciliation until it was the same as the stump-tails'. When de Waal removed the stump-tails, the rhesus kept on reconciling, like nice little stump-tailed ladies and gentlemen. It would be interesting to know what would have happened if the experiment had tried to impose rhesus mores on stump-tails. Would they be as quick to become morose grudge-holders, sabotaging one another's chances at tenure?

Another glimpse of rhesus potential comes from an experiment in which male rhesus were left alone with strange infants. Although males usually ignore infants, these males were solicitous, picking the babies up. But if there was a female in the cage, she would take charge of the baby's care (Don't hold him like that! Support his head! My God, give him to me!), and the same male would revert to standard hands-off behavior, appearing uncaring and unhelpful.

The increasing focus on peacemaking and cooperation bears on one of the most perennially vexing areas of evolutionary psychology, that of selfishness and altruism, especially as expounded in Richard Dawkins' 1976 book, *The Selfish Gene.* Unending confusion has been caused by Dawkins' use of the word "selfish," which switches between applying to individuals and meaning just what you think it means, you egotistical greed-head, and a very specific and different meaning invented by Dawkins and applying to genes—being an entity subject to natural selection.

Similarly "altruism" sometimes has an everyday meaning that includes giving to charity and snatching strangers out of the path of the runaway train, and sometimes means "self-destructive behavior performed for the benefit of others."

Partly as a result of this confusion, and partly because of the old fondness for uncomfortable truths, it has been widely announced that there is no such thing as altruism: The thing is genetically impossible. If you donate a kidney to a relative, you're only doing it because you share genes with them; if you give money to charity, you're only doing it to show off, so people will admire you and further your genes; and, worst of all, if you give money anonymously, you're only doing it to feel good about yourself, so you will more effectively be able to convince others that you are a decent creature, you big hypocrite. (See Robert Wright, *The Moral Animal*, for more on what self-deluded phonies we are, except maybe him.)

This confusion about altruism and selfishness is odd, considering that Dawkins can write with marvelous clarity. Why use metaphorical language that is so confusing and makes so many people upset? Well, it's certainly catchy. Perhaps there's also amusement in confronting people with the Uncomfortable Truth that everything they do is selfish. Andrew Brown, in *The Darwin Wars: How Stupid Genes Became Selfish Gods* (published in the UK, but not yet in the U.S.), suggests that when Dawkins was born, "The good fairy gave him good looks, intelligence, charm and a chair at Oxford . . . The bad fairy studied him for a while, and said 'Give him a gift for metaphor.' "

In the realm of altruism, another recent optimist is Lee Dugatkin, with *Cheating Monkeys and Citizen Bees: The Nature of Cooperation in Animals and Humans.* The bulk of the book analyzes "selfish" reasons for cooperating with relatives, partners (you brush flies off my face and I'll brush flies off yours), teams (lions hunting large prey together), and larger groups such as tribes and nations. It also touches lightly on Dugatkin's own work with guppies. (Don't scoff—these are *wild* guppies.)

Dugatkin is candid about his intellectual journey and how his views changed when he met his wife and when their son was born. Once, he says, "I was what I now refer to as a 'for the love of science' scientist . . . I actually held those scientists who directed their work toward helping people in contempt. I viewed them as intellectual prostitutes . . ." (He doesn't say he dashed about challenging people to face Uncomfortable Truths, but I bet he did.)

However, Dugatkin's marriage and fatherhood changed his perspective, and he is now determined to use his work for the benefit of all, and someday to share with his son "my thoughts on how ideas emerging from the study of animal cooperation might facilitate human sociality." (Why isn't this man roaming the streets hungry and homeless, hitting on women?)

Dugatkin charts evolutionary psychology along a political axis, arguing that liberals believe in the fundamental goodness of humans, and conservatives believe that humans are only good if taught to be so. This is nothing new—the social sciences and biological sciences, insofar as they describe humans, are perennially politicized. Natural selection has a history of popularity with some who equate their own success with Darwinian fitness. (De Waal gives the example of how this "convenient justification of disproportionate wealth in the hands of a happy few . . . led John D. Rockefeller to portray the expansion of a large business as 'merely the working-out of a law of nature and a law of God.' ")

This formulation makes Dugatkin a conservative, and he is urgent that people be taught to be good. In his final chapter, he describes himself as deeply religious. While humans are "God's crowning creation," animals have instructive worth. "If animals are here to help us, why should it not be the case that they can help us understand how to be more cooperative? . . . The beauty in this is that the animals need not even be cooperative themselves."

This pattern is common in works of evolutionary psychology: Hundreds of pages are devoted to the inescapability of "selfish" genetic mandates followed by a quick upward leap, appealing to the unique human intellect, unique human morality, or in this case, the unique love of God, to get us out of this deterministic hole.

Throughout recorded history people have denigrated animals as other and inferior, soulless, mindless instinct machines. Evolutionary psychology alarms people by putting us in the same despised category as animals—with the exception of those who go with the program, who understand and accept the uncomfortable truths of evolutionary psychology and thereby make an intellectual leap that places us above the genetic battleground. If we didn't place animals down so far, it wouldn't take such a vast leap to distinguish ourselves from them.

As befits a field that can't resist telling you what you're really doing when you think you're doing something else, the controversies of evolutionary psychology involve the motivations of scientists to a remarkable extent. Of course you think that, you're a liberal! Of course you think that, you're a man! Of course you think that, you want to drive me crazy! (Of course you think that, you're a human!) It's a bad sign for objectivity when knowing someone's political party, age or sex is likely to tell you where they stand on the issue of stump-tailed monkey destiny.

PETER GALISON

Einstein's Clocks:
The Place of Time

FROM *CRITICAL INQUIRY*

By examining in rich detail the periods in which scientists have made their discoveries, historians of science have shown how science can both sail with and beat against the intellectual currents of its day. Peter Galison, a distinguished historian of science at Harvard University, revisits the first years of the twentieth century, when a movement arose to normalize time zones and synchronize clocks across large areas. The quest for "universal time"was both an economic necessity and a metaphorical stand against anarchy. From his position at the Bern patent office, the young Einstein was well aware of this movement. Did it have a bearing on his formulation of relativity?

Einstein, 1933: "There are certain occupations, even in modern society, which entail living in isolation and do not require great physical or intellectual effort. Such occupations as the service of lighthouses and lightships come to mind."[1] Solitude, Einstein argued, would be perfect for the young scientist engaged with philosophical and mathematical problems. His own youth, we are tempted to speculate, might be thought of this way, the Bern patent office where he had earned a living seeming no more than a distant oceanic lightship. Consistent with this picture of otherworldliness, we have enshrined Einstein as the philosopher-scientist who, unmindful of the noise from his office work, rethought the foundations of his discipline and toppled the Newtonian absolutes of space and time.

Einstein's removal of these philosophical absolutes was more than a contribution to relativity; it has become a symbol of the overthrow of one philosophical epoch for another. To physicists such as Henri Poincaré, Hendrik Lorentz, and Max Abraham, Einstein's special relativity was startling, almost incomprehensible, because it began with basic assumptions about the behavior of clocks, rulers, and bodies in force-free motion—it began, in short, by assuming what these senior physicists had hoped to prove with starting assumptions about the structure of the electron, the nature of forces, and the dynamics of the ether. Soon a generation of physicists, including Werner Heisenberg and Niels Bohr, patterned its quantum epistemology around Einstein's quasi-operational definitions of space and time in terms of rulers and coordinated clocks. For the philosophers of the Vienna Circle, including Moritz Schlick, Rudolf Carnap, and Philipp Frank, Einstein's special relativity paper was also a turning point, an ever present banner to be flown for scientific philosophy.

For all these reasons, Einstein's 1905 "On the Electrodynamics of Moving Bodies" became the best-known physics paper of the twentieth century. Einstein's argument, as it is usually understood, departs so radically from the older, "practical" world of classical mechanics that the work has become a model of the revolutionary divide. Part philosophy and part physics, this rethinking of distant simultaneity has come to symbolize the irresolvable break of twentieth-century physics from that of the nineteenth. Recall the order of the argument. Einstein began with the claim that there was an asymmetry in the interpretation of Maxwell's equations, an asymmetry not present in the phenomena of nature (see figure on next page). A magnet approaching a coil produces a current indistinguishable from the current generated when a coil approaches a magnet. In Einstein's view this was a single phenomenon (coil and magnet approach and produce a current in the coil). But in their usual interpretation Maxwell's equations gave two different explanations of what was happening, depending on whether the coil or the magnet was in motion with respect to an all-pervasive ether. When the coil moved, charge within it experienced a force due to the static magnetic field; when the magnet moved, the changing magnetic field produced an electric field that drove the charge around the stationary coil. Einstein's goal was to produce a symmetric account, one that did not distinguish between the explanation given in the frame of reference of the coil and that given in the frame of reference of the magnet. The problem, as Ein-

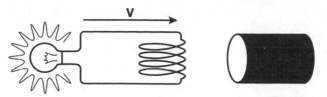

Static magnetic field forces moving charge around coil.

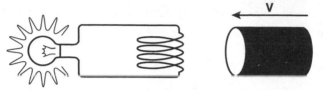

**Changing magnetic field makes electric field;
electric field drives static charge around coil.**

COIL MAGNET

stein diagnosed it, was that "insufficient consideration" had been paid to the circumstance that electrodynamics always depended on a view about *kinematics,* that is, about how clocks and rulers behaved in the absence of force.[2]

A coordinate system was, by Einstein's lights, a system of rigid measuring rods embodying Euclidean geometry and describable with ordinary Cartesian coordinates. So far, so good. Then comes the surprising part, the reanalysis of *time* that contemporaries like Hermann Minkowski saw as the crux of Einstein's argument.[3] As Einstein put it: "We have to take into account that all our judgments in which time plays a role are always judgments of *simultaneous events.* If, for instance, I say, 'That train arrives here at 7 o'clock,' I mean something like this: 'The pointing of the small hand of my watch to 7 and the arrival of the train are simultaneous events' " ("ZE," p. 893; "OE," p. 393).[4] For simultaneity *at a point,* there is no problem: if an event located immediately next to my watch (say, the train engine arriving next to me) happens just when the small hand of the watch reaches the seven, then those two events are said to be simultaneous. The difficulty, Einstein insists, comes when we have to link events at a distance: what would it mean to say two *distant* events are simultaneous?

In order to address this question, Einstein advances, in a seemingly philosophical vein, a thought experiment infinitely far from the exigencies of instruments, much less the daily considerations of patent office life. How, Einstein asks, ought we to *coordinate* our clocks? "We could in principle content ourselves to time events by using a clock-bearing observer located at the origin of the coordinate system, who coordinates the arrival of the light signal originating from the event to be timed . . . with the hands of his clock" ("ZE," p. 893; "OE, p. 393; trans. mod.). Alas, Einstein notes, because light travels at a finite speed, this system is not independent of the observer with the central clock. Two events judged simultaneous with respect to one origin will not be simultaneous if the origin is moved. This epistemic straw man will not tell good time (see figure on next page).

Young Einstein had a better system: let one observer at *A* send a light signal at noon to another at *B* a distance *d* away. *B* sets his clock to noon plus the time it takes a light signal to get to *B*, noon + *d/c*, where *c* is the speed of light. Continuing in this way, all other observers and their clocks are put in synchrony. With this system of coordination there is no special origin; there is no master clock. Here, so the account we have told ourselves goes, is the philosophical triumph of neo-Machian epistemic criticism over the fossilized absolutes of untouchable space and time. Einstein the philosopher-scientist has used thought experiments to vanquish unquestioned school dogma and a scientific-technical cadre too sophisticated to ask basic questions. But wait.

Let's go back to Einstein's train. You will recall that he wants to know what we mean by the arrival of a train in a station at seven o'clock. I have long followed Einstein himself in reading these remarks about trains and simultaneity as an instance of Einstein posing a question normally posed only "in early childhood," a matter that he, peculiarly, was still asking when he "was already grown up."[5] Such riddles about time and space appear, on this reading, to be so elementary, so basic, that they lay below the conscious awareness of the physics community. But was it, in fact, below the threshold of thought? Was no one else in 1904–05 in fact asking what it meant for a distant observer to know that a train was pulling into a station at seven o'clock? Was the idea of defining distant simultaneity such a philosophical reach?

This summer I was standing in a northern European train station, absent-mindedly staring at the turn-of-the-century clocks that lined the platform. They all read the same to the minute. Curious. Good clocks. But then I noticed that, as far as I could see, even the staccato motion of their second hands was in synchrony. These clocks were not simply running well, I thought, these clocks are coordinated. Einstein must have seen such coordinated clocks while he was grappling with his 1905 paper, trying to understand the meaning of distant simultaneity.

Already in the 1830s and 1840s Charles Wheatstone and Alexander Bain, both in England, and soon thereafter Mathias Hipp in Württemberg and a myriad of other inventors began constructing electrical distribution systems to

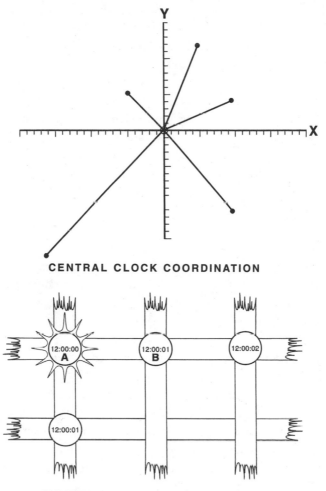

CENTRAL CLOCK COORDINATION

EINSTEIN'S CLOCK COORDINATION

bind distant clocks to a single central clock, called, variously, the *horloge-mère,* the *Primäre Normaluhr,* and the master clock.[6] In Germany Leipzig was the site of one of the first such electrically distributed time systems, followed by Frankfurt in 1859; Hipp (then director of a telegraph workshop) launched the Swiss effort at the Federal Palace in Bern, where a hundred clock faces began marching together in 1890. Geneva, Basel, Neuchâtel, and Zurich followed in quick succession, each with its own clock coordination system, and railroad lines—where clock coordination was vital—were soon rigged in Switzerland for coordinated time.[7]

Without coordinated times, cities, towns, and villages functioned on their own times, marking an individuality that remained unimportant before the railroad. In England during the 1830s London time led Reading time by four minutes, marched seven minutes and thirty seconds in advance of Cirencester, and chimed fourteen minutes before Bridgewater. If you wanted to display time on the front of a major building, you needed more than one clock. The Isle Tower in Geneva boasted three: the big clockface in the center showed Geneva time (about 10:13), the face on the left showed the Paris-based time used all along the track of the Paris-Lyon-Méditerranée railroad company (9:58), and the right-hand clock boasted Bern time—a handsome five minutes in advance (10:18). A few years later standardization entered the picture but proceeded line by line—every railroad company defined its own proper time and did so ceremoniously. In London

each morning an Admiralty messenger carried a watch bearing the correct time to the guard on the down Irish Mail leaving Euston for Holyhead. On arrival at Holyhead the time was passed on to officials on the Kingstown boat who carried it over to Dublin. On the return mail to Euston the watch was carried back to the Admiralty messenger at Euston once more.[8]

Germany, by most accounts the most advanced in its efforts to coordinate time, was still struggling with a hodgepodge of mechanical and electrical systems in 1891, when the aging Count Helmuth von Moltke came to speak, for the last time, to the Imperial German Parliament on 16 March; he died just over a month later.[9] As chief of the Prussian (and later German) general staff, von Moltke had dramatically reconceived the deployment of troops. Where earlier generations had sent armies over undependable roads, von Moltke exploited the railroads to supply, muster, and move troops over vast battlefronts. His successes with this strategy—in the Franco-Prussian War—no doubt brought the audience to attention when von Moltke pronounced on railroads, empire, and the military. In his scratchy voice he intoned:

Meine Herren, . . . I will not long detain you, as I am very hoarse, on which account I have to ask your indulgence.

That unity of time [Einheitszeit] is indispensable for the satisfactory operating of railways is universally recognized, and is not disputed. But, meine Herren, we have in Germany five different units of time. In north Germany, including Saxony, we reckon by Berlin time; in Bavaria, by that of Munich; in Württemberg, by that of Stuttgart; in Baden, by that of Carlsruhe; and on the Rhine Palatinate by that of Ludwigshafen. We have thus in Germany five zones, with all the drawbacks and disadvantages which result. These we have in our own fatherland, besides those we dread to meet at the French and Russian boundaries. This is, I may say, a ruin which has remained standing out of the once splintered condition of Germany, but which, since we have become an empire, it is proper should be done away with.

From the audience came the call "very true." Von Moltke went on to say that while this piecemeal ruin of time may only be an inconvenience for the traveler, it was an "actual difficulty of vital importance" for the railway business and, even worse, for the military. What, he asked, would happen in case of troop mobilization? There had to be a standard, one that would fall along the fifteenth meridian (about fifty miles east of the Brandenburg Gate) that would be the reference point. Local times within Germany would differ, but by a mere half hour or so on either extreme of the empire. "*Meine Herren,* unity of time merely for the railway does not set aside all the disadvantages which I have briefly mentioned; that will only be possible when we reach a unity of time reckoning for the whole of Germany, that is to say, when all local time is swept away."[10]

Von Moltke conceded that the public might dissent—wrongfully. But after some "careful consideration" the scientific men of the observatories would see things aright, and lend "their authority against this spirit of opposition." "*Meine Herren,* science desires much more than we do. She is not content with a German unity of time, or with that of middle Europe, but she is desirous of obtaining a world time, based upon the meridian of Greenwich, and certainly with full right from her standpoint, and with the end she has in view."[11] Farms and factory workers can offset their clock starting times as they wish. If a manufacturer wants his workers to start at the crack of dawn, then let him open the gates at 6:29 in March. Let the farmers follow the sun, let the schools and courts make do with their always loose schedules. Germany took action to extend its dominion of time, and much of Europe followed.

In Switzerland, the clock-world-famous Hipp (despite being arrested for consorting with anarchists) went on from his development of electrically maintained pendula to the practical deployment of a distribution system for time using low-tension circuits. Founded as a small telegraph and electrical apparatus factory in Neuchâtel, Hipp's company went from the establishment of the first network of public electric clocks in Geneva in 1861 to ever greater prominence; in 1889, it became A. de Peyer, A. Favarger & Cie. From 1889 to 1908 this concern extended the range of the mother clock beyond the dominion of railways to steeple clocks and even to the wake-up clocks inside hotels.[12]

With the march of time into every street, methods were needed to extend in-definitely the number of units that could be branched together—a flood of patents followed perfecting relays and signal amplifiers. The Bern urban time network was inaugurated in 1890; improvements, expansions, and new net-works sprouted throughout Switzerland. For not only was accurate, coordi-nated time important for European passenger railroads and the Prussian military, it was equally crucial for the dispersed Swiss clock-making industry that desperately needed means of consistent calibration.[13] But it was always practical and more than practical, at once material-economic necessity *and* cultural imaginary. Professor Wilhelm Förster of the Berlin Observatory, the observatory that set the Berlin master clock to the heavens, sniffed that any urban clock that did not guarantee time to the nearest minute was a machine "downright contemptuous of people."[14]

The burgeoning technology spun off patents in each sector of the network: patents on low-voltage generators, patents on electromagnetic receivers with all their escapements and armatures, patents on contact interrupters. Fairly typical of the kind of electrochronometric work blossoming in the years after 1900 was David Perret's novel receiver that would detect and use a direct-current chronometric signal to drive an oscillating armature, which was issued Swiss patent number 30351 at 5 P.M. on 12 March 1904. Or take Favarger's own receiver that did the opposite: it took an alternating current from the mother clock and turned it into the unidirectional motion of a toothed wheel. This patent—used widely—was submitted on 25 November 1902 and issued on 2 May 1905. There were patents on remote alarms, remote regulation of pen-dula, telephonic—even wireless—transmission of time; other patents arrived proposing clocks for railroad departures and arrivals, in addition to patents for clocks indicating time in other time zones.

All these chronometric patents—along with a great many others related to them—had to pass through the Swiss patent office in Bern, and no doubt many of them crossed Einstein's desk.[15] Einstein began work in the Bern patent office on 16 June 1902 as technical expert, third class, where he was chiefly charged with the evaluation of electromagnetic patents.[16] Standing at his wooden podium he, like the other twelve or so technical experts in the patent office, methodically went through each submission, searching for the princi-ples that lay at its core.[17] Einstein's expertise on electromechanical devices came in part from the family business. Indeed, Einstein's father, Hermann, and his uncle, Jakob Einstein, had built their enterprise out of his uncle's patents on sensitive, electrical clocklike devices for measuring electrical usage. One of J. Einstein & Cie.'s electrical meters, created by Jakob and Sebastian Kornprobst, was written up prominently in the report of the 1891 Frankfurt Electrotechnical Exhibit; just a few pages before is a mechanism (typical of the time) for mounting a backup mother clock to ensure the continued operation of a system of electrical clocks. So close were the electrical measuring systems and clockwork technologies that at least one of the Jakob Einstein-Sebastian Kornprobst patents explicitly registered its applicability to clockwork mechanisms more generally.[18]

Einstein's own later technical work (he held a great many patents) on what he called his *Machinchen* (a device for multiplying and measuring very small electrical charges) and his studies of the Einstein-de Haas effect (leading to his atomic theory of ferromagnetic atoms) were but two further examples of his particular interest in sensitive electromechanical devices. Electromagnetic clock coordination patents would have been right up his alley, as they centered around means of transforming small electrical currents into high-precision rotatory movements.[19]

Heading the patent office during Einstein's tenure was Friedrich Haller, a stern taskmaster to his underlings. Early on he reproached Einstein: "As a physicist you understand nothing about drawings. You have got to learn to grasp technical drawings and specifications before I can make you permanent."[20] In September 1903 Einstein received notice that his provisional appointment was permanent, though Haller was not ready to promote him, commenting that Einstein "should wait until he has fully mastered machine technology; he studied physics." That mastery came as Einstein plunged himself into the critical evaluation of the parade of patents that came before him. By April 1906 Einstein seems to have persuaded the authorities that, physics notwithstanding, he had indeed mastered the technology, and he was promoted to technical expert, second class. Haller now judged that Einstein "belongs among the most esteemed experts at the office."[21]

Einstein's window on the electrochronometric world came at a crucial time. For despite von Moltke's resounding support and the undamped enthusiasm of the advocates of one world time, Albert Favarger, one of Hipp's chief engineers and the man who effectively succeeded him at the helm of his company, was not at all content. At the 1900 Exposition Universelle, the International Congress on Chronometry met to discuss the status, inter alia, of clock coordination efforts.[22] At the outset of his speech to the congress Favarger rose to ask how it could be that the distribution of electrical time was running so distressingly far behind the related technologies of telegraphy or telephony. First, he suggested, there were technical difficulties; remotely coordinated clocks could rely on no obliging friend ("ami complaisant") to oversee and correct the least difficulty, whereas the steam engine, dynamo, or telegraph all seemed to run with constant human companionship ("SD," p. 198). Second, there was a technician gap: the best technical people were staffing power and communication devices, not time machines. And finally, he lamented, the public was not funding time distribution as it should. Such lagging boosterism baffled Favarger: "Could it be possible that we have not experienced the imperious, absolute, I would say collective need of time exactly, uniformly, and regularly distributed? . . . Here's a question that borders on impertinence when addressed to a late nineteenth-century public, laden with business and always rushed, a public that has made its own the famous adage: Time is money" ("SD," p. 199).

As far as Favarger was concerned, the sorry state of time distribution was all out of proportion with the exigencies of modern life. He insisted that humans needed a system of exactitude and universality correct to the nearest second. No old-fashioned mechanical, hydraulic, or pneumatic system would do—electricity was the key to the future, a future that would only come about properly if humankind broke with its mechanical clock past riven by anarchy, incoherence, and routinization. In its place, a world of electrocoordinated clocks must be based on a rational and methodical approach. As he put it,

> You don't have to run long errands through Paris to notice numerous clocks, both public and private, that disagree—which one is the biggest liar? In fact if even just one is lying one suspects the sincerity of them all. The public will only gain security when every single clock indicates unanimity at the same time at the same instant. ["SD," p. 200]

How could it be otherwise? Trains often barreled through the countryside in opposite directions on single tracks, and a shunting error of timekeeping could and did lead to calamity. Remote time regulation merging observatories, railroads, and telegraphy was all that stood between a smooth ride and smoking debris. Time went up for sale, and astronomers, telegraphers, and clockmakers all profited as they sent coordinated time down the railway lines. The first time zones were these long, thin territories carved by steel tracks.[23]

Favarger reminded the assembled exposition attendees that the speed of trains roaring through Europe was mounting—100, 150, even 200 kilometers per hour. Those running the trains and directing their movements—not to speak of the passengers trusting their lives to speeding carriages—had to have correct times. At fifty-five meters per second every tick counted, and the prevalent but obsolete mechanical systems of coordination were bound to be inferior. Only the electric, automatic system was truly appropriate: "The nonautomatic system, the most primitive yet the most widespread, is the direct cause of the time anarchy that we must escape" ("SD," p. 201).

Time anarchy. No doubt Favarger's reference was in part to the anarchism that had taken a powerful hold among the Jura watchmakers, as Pyotr Kropotkin testified in his *Memoirs of a Revolutionist*. To be sure, Kropotkin recorded, the theoretical issues raised by Bakunin and others against economic despotism were important,

> but the equalitarian relations which I found in the Jura Mountains, the independence of thought and expression which I saw developing in the workers, and their unlimited devotion to the cause appealed even more strongly to my feelings; and when I came away from the mountains, after a week's stay with the watchmakers, my views upon socialism were settled. I was an anarchist.[24]

Hipp himself had been arrested, but Favarger was clearly worried about more, about the broader disintegration of personal and societal regularity. Only electrical distribution of simultaneity could provide the "indefinite expansion of the time unification zone" ("SD," p. 202). Favarger's support for distant simultaneity was, all at once, political, profitable, and pragmatic.

Should we be able to escape from this dreaded "anarcho-clockism," we would have a chance to fill a great lacuna in our knowledge of the world. For, Favarger insisted, even as the International Bureau of Weights and Measures had begun to conquer the first two fundamental quantities—space and mass—time, the final frontier, remained unexplored (see "SD," p. 203). And the way to conquer time was to create an ever widening electrical network, enslaved to an observatory-linked mother clock that would drive relays multiplying its signals and send automatic clock resets into hotels, street corners, and steeples across continents. Tied in part to Favarger was a company that aimed to synchronize Bern's network. When, on 1 August 1890, Bern set the hands of its coordinated clocks in motion, the press hailed it as a "revolution in clocks."[25] It must not be forgotten that from many places in Bern you could clearly see several grand public clocks; when, that August, they all began running in step, time order became visible.

Swiss newspapers were not alone in seeing clock coordination as a matter of wide cultural import. To North American time booster Sandford Fleming and his allies in the 1890s the establishment of "universal" or "cosmic" time was both practical and more than practical—a boon to communication and transportation but also a "noiseless revolution" that would bring progress in all spheres of cultural and personal life.[26]

During the 1890s Einstein was not yet concerned about clocks at all; but as a young man of sixteen in 1895 he was already very much concerned with the nature of electromagnetic radiation. Even to his untutored imagination something was wrong with the customary conception of radiation as a wave in static, substantial ether. Suppose, he thought to himself, that one could catch up to a light wave—as classical physics might infer. Then, like a surfer riding an ocean wave train, he would see the electromagnetic field unfold before him oscillating in space but utterly unchanging in time. However, this corresponded to nothing ever observed.[27] Four years later Einstein was still agonizing over the nature of moving bodies and electrodynamics. To his beloved Mileva Marić he reiterated his sense that naive ether theories would simply have to go.

> D[ear] D[ollie],
> I returned the Helmholtz volume and am now rereading Hertz's propagation of electric force with great care because I didn't understand Helmholtz's treatise on the principle of least action in electrodynamics. I'm convinced more and more that the electrodynamics of moving bodies as it is presented today doesn't correspond to reality, and that it will be possible to present it in a simpler way. The introduction of the term "ether" into theories of electricity has led to the conception of a medium whose motion can be described, without, I believe, being able to ascribe physical meaning to it.[28]

Electricity and magnetism, Einstein concluded, would be definable as the motion of "true" electrical masses with physical reality through empty space. Ether, the centerpiece of nineteenth-century physical theories, was gone. So before Einstein set foot into the patent office crucial pieces of the relativity puzzle were in place: he knew Maxwell's equations, he was committed to a realistic picture of moving electrical charges, and he had dismissed the ether. But none of these considerations directly bore on the problem of how to treat time.

Meanwhile, some of the greatest physicists of the nineteenth century were beginning, out of desperation, to experiment with mathematical variations in the way the time variable t transformed in different reference frames. But all of them—Poincaré, Lorentz, Abraham—kept firmly to the notion of a true ether rest frame, and none of them accorded equal weight to these local times and the true (absolute) physical time of the ether rest system. Poincaré, Lorentz, and Abraham wanted to *begin* with special assumptions about the basic forces of nature, the forces that held together the atomic building blocks of interferometer arms, the forces that kept electrons from exploding because of their electrostatic self-repulsion. Out of such constructive, built-up theories of matter they sought to *deduce* kinematics—the behavior of matter in the absence of force. Einstein wanted none of that; he aimed for a theory that would start with simple physical principles, the way thermodynamics began with the conservation of energy and the increase of entropy. Poincaré, Lorentz, and Abraham were willing to make special assumptions about how an artificial, calculation-useful notion of time varied from frame to frame. None of them launched their detailed physics with a physical, principled set of assumptions about measured space and coordinated time.

So things stood in the years after Einstein arrived at the patent office in 1902.[29] In his patent work he had, in the instructions given by Haller, the directive to be critical at every stage: "When you pick up an application, think that anything the inventor says is wrong." To follow blindly would be to court disaster by following "the inventor's way of thinking, and that will prejudice you. You have to remain critically vigilant."[30] It was advice for patent work, but it applied as well to the ethereal realms of physics. For in the electrodynamics of moving bodies Einstein had a problem that had troubled him on and off for some seven years, a problem that was with increasing force agonizing the leading physicists of the day. Meanwhile, all around him, literally, was the burgeoning fascination with electrocoordinated time. Every day Einstein took the short stroll from his house, left down the Kramgasse, to the patent office; every day he must have seen the great clock towers that presided over Bern with their coordinated clocks, and the myriad of street clocks branched proudly to the central telegraph office. After all, he had to walk under one of the most famous of them, the Zeitglockenturm, and by many others. Sometime in the middle of May 1905—and we note that Einstein moved to the edge of Bern's unified time zone on 15 May—he and his closest friend, Michel Besso, cornered the electromagnetism problem from every side. "Then," Einstein recalled, "suddenly I understood where the key to this problem lay." He skipped his greetings the next day when he met Besso: " 'Thank you; I've completely solved the problem.' An analysis of the concept of time was my solution. Time cannot be absolutely defined, and there is an inseparable relation between time and signal velocity."[31] Pointing up at a Bern clock tower—one of the famous synchronized clocks in Bern—and then to a clock tower in nearby Muri (not yet linked to the Bern mother clock), Einstein laid out his synchronization of clocks.[32]

Within a few days Einstein sent off a letter to his friend Conrad Habicht imploring him to send a copy of his dissertation and promising four new papers in return. "The fourth paper is only a rough draft at this point, and is an electrodynamics of moving bodies which employs a modification of the theory of space and time; the purely kinematic part of this paper [beginning with the new definitions of time synchronization] will surely interest you."[33] Ten years of thought had gone into this problem, but time synchronization was the final, crowning step in the development of special relativity.

In this light, Einstein's paper, completed by the end of June 1905, can now be read in a very different fashion. Instead of a pure "Einstein philosopher-scientist" merely earning his keep in the patent office, we can see him also as "Einstein patent-officer-scientist" refracting the underlying metaphysics of his relativity theory through some of the most symbolized mechanisms of modernity. The train arrives in the station at seven o'clock, as before, but now it is not just Einstein who is worried about what this means in terms of distant simultaneity. No, determining train arrival times using electromagnetically coordinated clocks was *precisely* the technological issue that had been racking Europe. Patents now raced through the system, improving the electrical pendula, altering the receivers, introducing new relays, and expanding system capacity. Time coordination in the central Europe of 1902–05 was no arcane subject; it was front and center for the clock industry, the military, and the railroad *as well as* a symbol of the interconnected, sped-up world of modernity.

By addressing the problem of distant simultaneity, Einstein was engaging a powerful and highly visible new technology that conventionalized simultaneity, first to synchronize train lines and to set longitude, and then to fix time zones. It was in this world that Einstein brought a conventional basis into his vision of a principled physics. A trace of the existing time coordination system is there to see in the 1905 paper itself. Reconsider the scheme of coordination that Einstein *explicitly refused* to accept: an observer equipped with a clock at the center of the coordinate system. That master clock bolted to space position (0,0,0) determines simultaneity when electromagnetic signals from distant points arrive there at the same local time. But now this standard centered system no longer appears as an abstract straw man. This branching, radial clock coordination structure, visible in wires, generators, and clocks, displayed in book after book on timekeeping, *was precisely that of the European system of the mother clock along with its secondary and tertiary dependents*. When center-issued signals arrived at distant points, be they in the next room or a hundred kilometers away, they were *defined* as simultaneous; on that basis trains were run, troops were rousted, and telegraph messages were sent. It is even in this period that preparations were being made to send the time coordination signal out by radio waves. There was an intense burst of activity on such radio coordination systems in 1904 both in Switzerland and in France as workers tested, developed, and began deploying the new radio time system. The director of *La Nature* himself took up his pen to record new developments in the distribution of time by wireless methods. Reporting on experiments conducted at the Paris Observatory, he noted that with the aid of a chronograph distant synchronization now appeared to be possible to within two or three hundredths of a second—and the wireless technologies promised to distribute time everywhere throughout Paris and its surrounding suburbs. Not only would scientific goals be advanced, such as the determination of longitude, but freed from the constraints of physical wire, time could be broadcast out to boats at sea and even into the ordinary household.[34] By 1905 the American navy was using radio-controlled clocks, and by 1910 the Eiffel Tower station was fixing the hands of clocks across Europe. According to one leading radio time worker in 1911, planning for radio simultaneity had begun with radio itself, presumably sometime around 1901.[35] But whether by telegraph line or by wireless, centralized time distribution was the temporal-physical glory of the unified German empire that von Moltke wanted, made corporeal through the grand *Primäre Normaluhr* at the Silesischer Bahnhof in Berlin or the baroque and elegant *horloge-mère* of Neuchâtel.

For the telegraphers, geodesists, and astronomers, Einstein's novel time coordination scheme could clearly be understood in terms of the already extant clock coordination methods. At the École Professionelle Supérieure des Postes et Télégraphes, on 19 November 1921, when Léon Bloch sought to explain the meaning of time, he turned his audience's attention to the actual and widespread technology that they would have known like the backs of their hands:

> What do we call time on the surface of the earth? Take a clock that gives astronomical time—the mother pendulum of the Observatory of Paris,—and transmit that time by wireless to distant sites. In what does this transmission consist? It consists of noting at the two stations that need synchronization, the passage of a common luminous or hertzian signal.[36]

At least at the Postes et Télégraphes, relativity was understood through their real infrastructure of coordinated clocks. Yet more than the mere invocation of clock synchronization is involved. By 1924, and very probably some time before then, clock coordinators had begun (following Einstein) to take account of the finite velocity of radio waves. Einstein's universal time machine had rapidly drawn from the technological world, reshaped the physical-metaphysical one, and now begun to reconfigure machinery.[37]

Indeed, given the physical—and cultural—impact of clock coordination, and its place in such a variety of settings, one might well ask if before 1905 anyone else besides Einstein was worried about defining time through a rigorous synchronization that took on board the finite velocity of light. Astonishingly enough there was one other person, someone who perhaps quite reasonably had been a member of the French Bureau of Longitudes since 4 January 1893, ascending to its presidency in 1899 (and in 1909), longitude determination having been for centuries the domain where clock coordination had been critical. To boot he was professor at the École Professionelle Supérieure des Postes et Télégraphes from 4 July 1902; when electrocoordinated clocks came on the scene it was Postes et Télégraphes that took charge. That other was, of course, Poincaré.[38] He—like Einstein—introduced clocks coordinated by the exchange of a luminous signal.

Poincaré's first exploration of simultaneity came in 1898, when he argued that simultaneity was not an absolute concept, insisting that we have no direct intuition to any such notion. What we do have are certain rules, rules that we must invoke in order to do the quite concrete technical work of, for example, longitude determination. "When sailors or geographers determine a longitude they have precisely the problem to solve that we have been treating here; they must, without being at Paris, determine the time at Paris. How do they do it?" They could move a clock (with all the problems that attend such an effort), they could both refer to an astronomical event, or, finally, "they could make use of the telegraph. It is clear first of all that the reception of a signal in Berlin, for example, is posterior to the sending of that same signal from Paris."[39] This is not a purely hypothetical example. Commentators have often treated Poincaré's telegraphy reference as if it were an imaginary problem, an instance of abstract philosophical rumination. It was not. By 1898 an actual system of clock coordination existed in the service of longitude determination. Indeed, the geodetic and meteorological demand for coordinated time (to determine longitude) had joined with the exigencies of railroad economy and safety to launch the project of time zones.[40]

Poincaré, as noted, was a senior member of the Bureau of Longitudes in 1898, and his reference was not to telegraphy as an abstract instance of finite signal propagation but rather, explicitly, as a means of clock-coordinated longitude work:

> In general, the duration of the transmission [from Paris to Berlin] is neglected and the two events are regarded as simultaneous. But, to be rigorous, a little correction would still have to be made by a complicated calculation.

In practice such a correction is not made because it would be smaller than observational errors, but its theoretical necessity is not diminished from our point of view, which is that of providing a rigorous definition [of simultaneity]. ["M," p. 12; "MT," p. 35; trans. mod.]

"Theoretically necessary" was the recognition that simultaneity was *always* conventional. And in this instance, at least, Poincaré's conventionalism was tied more than homonymically to the grubby world of real conventions that were, at precisely this conjuncture, setting time zone simultaneity, railway simultaneity, and national time unification.

Sometime after the spring of 1902 Einstein might have read Poincaré's "The Measure of Time." We know from Einstein's friend Maurice Solovine that their little discussion group (grandly titled the Olympia Academy) definitely read a later Poincaré work, *Science and Hypothesis*, that cited the 1898 work.[41] Critical as Poincaré was of any attempt to pretend that time-coordinating conventions were intuitive absolutes, on *practical* grounds, as he suggested in the remarks above, he did not militate for the abandonment of Newtonian kinematics. Ordinary rules of simultaneity, "the fruit of an unconscious opportunism," were "not imposed upon us and we might amuse ourselves in inventing others; but they could not be cast aside without greatly complicating the enunciation of the laws of physics, mechanics and astronomy" ("M," p. 13; "MT," p. 36). Corrections to the Newtonian world, Poincaré believed, would be small and complicated, so "theoretical necessity" would be trumped by the demands of simplicity. Einstein looked at distant simultaneity almost exactly as had Poincaré. But where Poincaré saw the new light-signal synchronization as leading to inevitable complexity, Einstein saw it as the harbinger of a vastly *simpler* physics.[42]

The turn-of-the-century Euro-American world was crisscrossed with overlapping networks of coordination: webs of train tracks, telegraph lines, meteorological networks, longitude surveys under the watchful, increasingly universal, clock system. In this context, the clock coordination system introduced by Einstein was, in a nontrivial sense, a world-machine, a vast, at first only imagined network of clocks. At the risk of seeming self-contradictory, there is a sense in which Einstein's special theory of relativity was a machine, an imaginative one, to be sure, but one built nonetheless on a real skein of wires that synchronized time machines by the exchange of electromagnetic signals.

Such a *technological* reading of this most *theoretical* paper suggests one final observation. It has long struck scholars that the style of "On the Electrodynamics of Moving Bodies" does not even look like that of an ordinary physics paper. There are essentially no footnotes, very few equations, no mention of new experimental results, and a lot of banter about simple physical processes that seem far removed from the frontiers of science.[43] Pick up a typical paper from the *Annalen der Physik* and a very different form appears in nearly every article: they typically begin with an experimental problem, a calculational correction; they are filled with references to other papers. But read Einstein's paper through the eyes of the patent world and suddenly it looks not so strange—at least in style. As one recent author has commented, patents are *precisely* characterized by their refusal to lodge themselves among other patents through footnotes—that would compromise the entrepreneurial advantage the author seeks. The simplistic banter is not unusual; patents are in fact written for a "person skilled in the art" (as patent language had it), not specialist readers.[44] And the story told by the author aims to describe a procedure rather than point the way in a long train of work.

Given the embedding of this scientific-technological intervention in the wider world of time coordination another puzzle arises. In a sense Einstein—the same Einstein who at sixteen had abandoned his German citizenship and who over a lifetime lambasted the "herd life" of "the military system"—had, ironically, at age twenty-six, completed von Moltke's project.[45] Time was identified with timekeeping, and *Einheitszeit* became the technopolitical endpoint of establishing, procedurally, distant simultaneity in an ever expanding domain. Einstein's clock synchronization system, like its predecessors, reduced time to procedural synchronicity, tying clocks together by electromagnetic signals. And in Einstein's scheme clock unity extended beyond city, country, and empire, beyond continent, indeed, beyond the world, to the infinite, now pseudo-Cartesian, universe as a whole.

But the irony inverts. For while Einstein's clock coordination procedure built on at least fifteen years of intense efforts towards electromagnetic time unification, he had, devastatingly, removed one crucial element of von Moltke's vision. There was, in Einstein's infinite clock machine, no *Primäre Normaluhr*, no *horloge-mère*, no master clock. His was a coordinated system of infinite spatiotemporal extent, but its infinity was without center—no Silesischer Bahnhof linked upwards through the Berlin observatory to the heavens and downwards to the edges of empire. By infinitely extending a time unity that had originally been grounded in the imperatives of German national unity, Einstein had both completed and subverted the project. He had opened the zone of unification, but in the process he not only removed Berlin as the *Zeitzentrum* but also designed a machine that upended the very category of metaphysical centrality. With time coordination now defined by the exchange of electromagnetic signals, Einstein could finish his description of the electromagnetic theory of moving bodies with no spatial or temporal reference to any specially picked out ether rest frame. He had a theory of relativity in which the asymmetry between reference frames was gone.

TIMES CHANGE. Einstein left the Bern patent office in 1909 for the University of Zurich, then went to Prague, and eventually in 1914 took up his post at the University of Berlin. After World War I Favarger, that avatar of Swiss chronometric unity, published his technical 550-page third edition of his treatise on electrical timekeeping, framing it, once again, in broadly cultural terms. The Great War, he argued, had contributed powerful technical developments, but it also destroyed a great part of the human wealth that sustained peace had created. The world remaining was by contrast "a heap of ruins, of miseries, and of suffering."[46] To exit this disaster, at least materially, work would save humanity, and work, mechanics teaches us, is a product of time and force. Time, he goes on to say, "cannot be defined in substance; it is, metaphysically speaking, as mysterious as matter and space." (Even stolid Swiss clockmakers were driven to metaphysics by time.) All the activities of man, whether conscious or unconscious—sleeping, eating, meditating, or playing—take place in time; without order, without specified plans, we risk falling into anarchy—into "physical, intellectual and moral misery." The remedy: the precise measurement and determination of time with the rigor of an astronomical observatory. But measured time cannot remain in the astronomers' redoubt; time rigor must be distributed electrically to anyone who wants or needs it: "we must, in a word, popularize it, we must *democratize* time" for people to live and prosper. We must make every man *"maître du temps,* master not only of the hour but also of the minute, the second, and even in special cases the tenth, the hundredth, the thousandth, the millionth of a second."[47] Distributed, coordinated time was more than money for Favarger; it was each person's access to orderliness, interior and exterior. Throughout the late nineteenth and early twentieth centuries, coordinated clocks were never just gears and magnets.

My hope in exploring the material culture of clock coordination is to set Einstein's place in a universe of meaning that crossed mechanisms and metaphysics. More generally, my hope here—and elsewhere[48]—for studies in the material culture of science is to avoid two equally problematic positions on the relation of things to thoughts. On one side we have a long tradition of unregenerate materialism or empiricism, a view that ideas emerge causally and univalently from the disposition of objects and the impressions they make upon us. In the history of physics, empiricism of a specifically logical positivist sort directly and unambiguously shaped an inductive, observation-centered account of scientific development, codified in the *Harvard Case Histories in Experimental Science* but present throughout the 1950s and into the 1960s. In that frame, theory, and the philosophy with which it was associated, was an always provisional addition, not the bulwark of science. Einstein here appears as having taken the inexorable next step in an inductive process that gradually drove out the ether: the ether couldn't be measured to first order in the ratio of velocity to the speed of light (v/c), it wasn't there to second order in v/c, and therefore (so the argument went), Einstein concluded that the ether was superfluous.[49] No doubt there is much to be said for this experiment-grounded Einstein—his fascination with the detailed conduct of the fast electron experiments and his gyrocompass work at the Physikalisch-Technische Reichsanstalt reveal a theorist with a clear sense of laboratory procedure and the operation of machines. Things structured thoughts.

On the other side, and characteristic of the inverse project, was the antipositivist movement of the 1960s and 1970s that largely aimed at reversing the previous generation's epistemic order: programmes, paradigms, and conceptual schemes came first, and these reshaped experiments and instruments all the way down. Now thoughts fully structured things. Einstein on the antipositivist screen appears as the philosophical innovator who dispensed with the material world altogether in a sustained drive for symmetry, principles, and operational definitions. Much truth here, too; from the antipositivist reaction we learned to be sensitive to those moments when Einstein was chary of experimental results, dubious, for example, about supposed laboratory refutations of special relativity and about nominal astronomical contradictions with the general theory.

Granting both historiographical traditions their due, I am not proposing to split the difference, and I am certainly not advocating a technological reductivism. Instead, it seems to me that we have, in the form of a philosophically informed and historicized material culture, a way out of this binary oscillation between a historiography of implicit idealism or implicit materialism. Work on telegraphs, steam engines, scientific instruments, and astronomical observation over the last years has set questions that refuse the untenable either/or of things and thoughts.[50] In each instance we can explore the philosophical issues engaged with historically specific values and symbols.

When Einstein came to the Bern patent office in 1902 he entered into a world in which the triumph of the electrical over the mechanical was already symbolically wired to dreams of modernity. He found a world in which clock coordination was a practical problem (trains, troops, and telegraphs) demanding workable, patentable solutions in exactly his area of greatest concern and professional occupation: precision electromechanical instrumentation. The patent office was anything but a deep-sea lightship. No, the office was a grandstand seat for the great parade of modern technologies. And as coordinated clocks went by, they weren't traveling alone; the network of electrical chronocoordination signified political, cultural, and technical unity all at once. Einstein seized on this new, conventional simultaneity machine and installed it at the principled beginning of his new physics. In a certain sense he had completed the grand time coordination project of the nineteenth century, but by eliminating the master clock and raising the conventionally set time to a *physical principle*, he had launched a distinctively modern twentieth-century physics of relativity.

Hypersymbolized—by which I mean many competing interpretations were in play—the regulated coordination of *Einheitszeit* meant, alternately, imperial empire, democracy, world citizenship, and antianarchism. What they held in common was a sense that each clock signified the individual and that clock coordination came to stand in for a logic of linkage among people and peoples. As such, the project for country or worldwide regulation elicited specific conditions of possible technocultural moves—that is, moves that would at once bear both scientific-technological and cultural significance.

It has become a commonplace over the last thirty years to pit bottom-up against top-down explanations. Neither will do. Borrowing a medieval saying aimed at capturing the links between alchemy and astronomy, we might put it this way: In looking down—to the electromagnetically regulated clock networks—we see up—to images of empire, metaphysics, and civil society. In looking up—to the metaphysics of Einstein's operationalized distant simultaneity, to the shifting culture of space, time, and motion—we see down—to the wires, gears, and pulses passing through the Bern patent office. We find metaphysics in machines, and machines in metaphysics.

Notes

Unless otherwise indicated, all translations are my own.

1. Albert Einstein, speech delivered at the Royal Albert Hall, London, 3 Oct. 1933, in *Einstein on Peace*, ed. Otto Nathan and Heinz Norden (New York, 1960), p. 238.

2. Einstein, "Zur Elektrodynamik bewegter Körper," *Annalen der Physik* 17 (1905): 892, hereafter abbreviated "ZE"; trans. Arthur I. Miller, under the title "On the Electrodynamics of Moving Bodies," appendix in *Albert Einstein's Special Theory of Relativity:*

Emergence (1905) and Early Interpretation (1905–1911) (Reading, Mass., 1981), p. 393, hereafter abbreviated "OE."

3. See Peter L. Galison, "Minkowski's Space-Time: From Visual Thinking to the Absolute World," *Historical Studies in the Physical Sciences* 10 (1979): 85–121.

4. We have all learned to read Einstein's papers in no small measure through the extensive work by Gerald Holton, *Thematic Origins of Scientific Thought: Kepler to Einstein* (Cambridge, Mass., 1973). I also find very helpful the more recent work by Abraham Pais, *Subtle Is the Lord* (Oxford, 1982), and Andrew Warwick, "On the Role of the FitzGerald-Lorentz Contraction Hypothesis in the Development of Joseph Larmor's Electronic Theory of Matter," *Archive for History of Exact Sciences* 43, no. 1 (1991): 29–91, "Cambridge Mathematics and Cavendish Physics: Cunningham, Campbell, and Einstein's Relativity, 1905–1911, Part I: The Uses of Theory," *Studies in the History and Philosophy of Science* 23 (Dec. 1992): 625–56, and "Cambridge Mathematics and Cavendish Physics: Cunningham, Campbell, and Einstein's Relativity, 1905–1911, Part II: Comparing Traditions in Cambridge Physics," *Studies in the History and Philosophy of Science* 24 (Mar. 1993): 1–25; Richard Staley, "On the Histories of Relativity: The Propagation and Elaboration of Relativity Theory in Participant Histories in Germany, 1905–11," *Isis* 89 (June 1998): 263–99; and Albrecht Fölsing, *Albert Einstein: A Biography,* trans. Ewald Osers (New York, 1997), p. 155.

5. Quoted in *Helle-Zeit-dunkle Zeit: In Memoriam Albert Einstein,* ed. Carl Seelig (Zurich, 1956), p. 71; trans. in *The Quotable Einstein,* ed. Alice Calaprice (Princeton, N.J., 1996), p. 182.

6. Remotely set clocks were discussed by, among others, Charles Wheatstone and William Cooke, the Scottish clockmaker Alexander Bain, and the American inventor Samuel F. B. Morse. For Wheatstone, Cooke, and Morse, clock coordination came out of their work on telegraphy. See Kenneth F. Welch, *Time Measurement: An Introductory History* (Newton Abbot, 1972), pp. 71–72.

7. For discussions of the extensive work on clock coordination before 1900, see, for example, the series of articles by A. Favarger, "L'Électricité et ses applications à la chronométrie," *Journal suisse d'horlogerie* 9 (Sept. 1884–June 1885), esp. pp. 153–58; Favarger, "Les Horloges électriques," in *Histoire de la pendulerie neuchâteloise,* ed. Alfred Chapuis (Paris, 1917), pp. 399–420; Friedrich Anton Leopold Ambronn, *Handbuch der astronomischen Instrumentenkunde,* 2 vols. (Berlin, 1899), esp. 1:183–87. On the expansion of the Bern network, see Gesellschaft für elektrische Uhren in Bern, *Jahresberichte* (1890–1910), Stadtarchiv Bern.

8. Philip S. Bagwell, *The Transport Revolution from 1770* (London, 1974), p. 125; quoted in Wolfgang Schivelbusch, *The Railway Journey: Trains and Travel in the Nineteenth Century,* trans. Anselm Hollo (New York, 1980), pp. 48–50.

9. The establishment of uniform time is discussed in Stephen Kern, *The Culture of Time and Space, 1880–1918* (Cambridge, Mass., 1983), pp. 11–14, and in Derek Howse, *Greenwich Time and the Discovery of Longitude* (Oxford, 1980), pp. 119–20. Simon Schaffer uses H. G. Wells's time machine as a guide through the turn-of-the-century intersection of the mechanized workplace and the literary and scientific engagement with time in "Time Machines" (unpublished manuscript, University of Cambridge).

10. Helmuth von Moltke, "Dritte Berathung des Reichshaushaltsetats: Reichseisenbahnamt Einheitszeit," *Gesammelte Schriften und Denkwürdigkeiten des General-Feldmarschalls Grafen Helmuth von Moltke,* 8 vols. (Berlin, 1891–93), 7:38–39, 39, 40; trans. Sandford Fleming, under the title "General von Moltke on Time Reform," in *Documents*

Relating to the Fixing of a Standard of Time and the Legalization Thereof, Canada Parliament Session 1891, no. 8, pp. 25, 25, 26; trans. mod.

11. Von Moltke, "Dritte Berathung des Reichshaushaltsetats: Reichseisenbahnamt Einheitszeit," p. 40; Fleming, "General von Moltke on Time Reform," p. 26; trans. mod.

12. For biographical details on Hipp, see Aymon de Mestral, *Mathias Hipp, 1813–1893; Jean-Jacques Kohler, 1860–1930; Eugène Faillettaz, 1873–1943; Jean Landry, 1875–1940* (Zurich, 1960), pp. 9–34. David S. Landes's work, *Revolution in Time: Clocks and the Making of the Modern World* (Cambridge, Mass., 1983), pp. 237–337, is excellent on the Swiss watch industry, though he is not focussed here on networks but rather on clock production.

13. See Favarger, *L'Électricité et ses applications à la chronométrie*, 3d ed. (Neuchâtel, 1924), pp. 408–9.

14. Gerhard Dohrn-van Rossum, *History of the Hour: Clocks and Modern Temporal Orders*, trans. Thomas Dunlap (Chicago, 1996), p. 350. See also Ulla Merle, "Tempo! Tempo! Die Industrialisierung der Zeit im 19. Jahrhundert," in *Uhrzeiten: Die Geschichte der Uhr und ihres Gebrauches*, ed. Igor A. Jenzen (Frankfurt am Main, 1989), pp. 166–78.

15. Hundreds of relevant patents are listed in the *Journal suisse d'horlogerie* during the relevant years (1902–5). Sadly, the Swiss patent office dutifully destroyed all papers processed by Einstein eighteen years after their creation; this was standard procedure on patent opinions, and even Einstein's fame led to no exception. See Fölsing, *Albert Einstein*, p. 104.

16. The most detailed linkage between Einstein's patent work and his scientific work is on gyromagnetic compasses and Einstein's production of the Einstein-de Haas effect. See Galison, *How Experiments End* (Chicago, 1987), chap. 2; in addition, see Thomas Hughes, "Einstein, Inventors, and Invention," *Science in Context* 6 (Spring 1993): 25–42, and Lewis Pyenson, *The Young Einstein: The Advent of Relativity* (Bristol, 1985). On Einstein's assignment to evaluate electrical patents, see Max Flückiger, *Albert Einstein in Bern: Das Ringen um ein neues Weltbild: Eine dokumentarische Darstellung über den Aufstieg eines Genies* (Bern, 1974), p. 62.

17. See Flückiger, *Albert Einstein in Bern*, p. 66.

18. See J. Einstein & Cie. and Sebastian Kornprobst, "Vorrichtung zur Umwandlung der ungleichmässigen Zeigerausschläge von Elektrizitäts-Messern in eine gleichmässige, gradlinige Bewegung," Kaiserliches Patentamt no. 53546, 26 Feb. 1890; "Neuerung an elektrischen Mess- und Anzeigervorrichtungen," Kaiserliches Patentamt no. 53846, 21 Nov. 1889; "Federndes Reibrad," Kaiserliches Patentamt no. 60361, 23 Feb. 1890; and "Elektrizitätszähler der Firma J. Einstein & Cie., München (System Kornprobst)," *Offizielle Zeitung der Internationalen Elektrotechnischen Ausstellung*, no. 28 (Oct. 1891): 949. See also Viktor Yakovlevitch Frenkel and Boris Efimovitch Yavelov, *Einshtein: Izobreteniia i eksperiment* (Einstein: invention and experiment) (Moscow, 1990), pp. 75–79, and Pyenson, *The Young Einstein*, pp. 39–53.

19. On the "little machine" *(Machinchen)*, see John Stachel et al., "Einstein's 'Machinchen' for the Measurement of Small Quantities of Electricity," editorial note in *The Swiss Years: Correspondence, 1902–1914*, vol. 5 of *The Collected Papers of Albert Einstein*, trans. Anna Beck, ed. Stachel et al. (Princeton, N.J., 1995), pp. 51–55; on the Einstein-de Haas effect, see Galison, *How Experiments End*, chap. 2, and Frenkel and Yavelov, *Einstein*, chap. 4.

20. Quoted in Flückiger, *Albert Einstein in Bern*, p. 58.

21. Quoted in Pais, *Subtle Is the Lord*, pp. 47–48.

22. See Favarger, "Sur la distribution de l'heure civile," in Congrès International de Chronométrie, *Comptes rendues des travaux, procès-verbaux, rapports, et mémoires*, ed. E. Fichot and P. de Vanssay (Paris, 1902), pp. 198–203; hereafter abbreviated "SD."

23. See Carlene Stephens, " 'The Most Reliable Time': William Bond, the New England Railroads, and Time Awareness in Nineteenth-Century America," *Technology and Culture* 30 (Jan. 1989): 1–24 and "Before Standard Time: Distributing Time in Nineteenth-Century America," *Vistas in Astronomy* 28, pts. 1–2 (1985); 114–15.

24. Peter Kropotkin, *Memoirs of a Revolutionist*, trans. pub. (Montreal, 1989), p. 267.

25. Quoted in Jakob Messerli, *Gleichmässig pünktlich schnell: Zeiteinteilung und Zeitgebrauch in der Schweiz im 19. Jahrhundert* (Zurich, 1995), p. 126.

26. Fleming, *Time-Reckoning for the Twentieth Century* (Washington, D.C., 1889), p. 357. See Ian R. Bartky, "The Adoption of Standard Time," *Technology and Culture* 30 (Jan. 1989): 41 for links of Fleming to Cleveland Abbe and other meteorologists.

27. See Einstein, "Autobiographical Notes," in *Albert Einstein: Philosopher-Scientist*, ed. Paul Arthur Schilpp, 3d ed. (La Salle, Ill., 1970), p. 53.

28. Einstein to Mileva Marić, 10? Aug. 1899, in Einstein and Marić, *The Love Letters*, trans. Shawn Smith, ed. Jürgen Renn and Robert Schulmann (Princeton, N.J., 1992), p. 10. On Einstein's specific knowledge of aspects of electrodynamics, see Holton, "Influences on Einstein's Early Work," *Thematic Origins of Scientific Thought*, and Miller, *Albert Einstein's Special Theory of Relativity*.

29. Here is not the place to offer a reconstruction of all aspects of Einstein's path to special relativity. The reader is referred to an excellent short synthesis in Stachel et al., "Einstein on the Special Theory of Relativity," editorial note in *The Swiss Years: Writings, 1900–1909*, vol. 2 of *The Collected Papers of Albert Einstein*, ed. Stachel et al. (Princeton, N.J., 1989), pp. 253–74, esp. pp. 264–65, which argues that the rough sequence of Einstein's work was (1) conviction that only relative motion of ponderable bodies was significant; (2) abandonment of Lorentz's assignment of physical significance to absolute motion; (3) exploration of alternative electrodynamics justifying emission hypothesis of light relative to source; (4) abandonment of this alternative electrodynamics as Einstein assumes velocity of light independent of the velocity of the source; (5) critique of the usual conception of temporal and spatial intervals, and especially of distant simultaneity; and (6) physical definition of simultaneity and the construction of a new kinematic theory. Here my focus is on (5), the introduction of a conventional notion of distant simultaneity.

30. Quoted in Flückiger, *Albert Einstein in Bern*, p. 58.

31. Einstein, "How I Created the Theory of Relativity," lecture, Kyoto, 14 Dec. 1922, trans. Yoshimasa A. Ono, *Physics Today* 35 (Aug. 1982): 46.

32. See Josef Sauter, "Comment j'ai appris à connaître Einstein," in Flückiger, *Albert Einstein in Bern*, p. 156, and Fölsing, *Albert Einstein*, p. 155.

33. Einstein to Conrad Habicht, Bern [18 or 25 May 1905], *The Swiss Years: Correspondence, 1902–1914*, p. 20.

34. See Henri de Parville, "Distribution de l'heure par télégraphie sans fil," *La Nature*, 30 July 1904, pp. 129–30. Experiments were conducted by G. Bigourdan, astronomer at the Paris Observatory, and presented to the Académie des Sciences on 27 June 1904; these results were printed in the *Comptes rendues de l'Académie* and quoted *in extenso* (along with the work of others including the director of the observatory at Neuchâtel) in "La Télégraphie sans fil et la distribution de l'heure," *Journal suisse d'horlogerie* 29 (Sept. 1904): 81–83.

35. On wireless time setting, see, for example, Joseph Roussel, *Le Premier Livre de l'a-*

mateur de T.S.F. (Paris, 1922), esp. pp. 150–52. Julien Auguste Boulanger and Gustave Auguste Ferrié, *La Télégraphie sans fil et les ondes électriques*, 7th ed. (Paris, 1909) dates the Eiffel Tower radio station to 1903. Ferrié, "Sur quelques nouvelles applications de la télégraphie sans fil," *Journal de Physique*, 5th ser., 1 (1911): 178–89, esp. p. 178, indicates that planning for wireless time coordination began at the start of work on wireless; Edmond Rothé, *Les Applications de la télégraphie sans fil: Traité pratique pour la réception des signaux horaires* (Paris, 1913) discusses the details of radio-communicated time coordination procedure.

36. Léon Bloch, *Le Principe de la relativité et la théorie d'Einstein* (Paris, 1922), pp. 15–16. Dominique Pestre characterizes Bloch (and his brother) as physicists who were unusual for their time in France by virtue of writing textbooks that looked positively on the new physics of the early twentieth century, and who characteristically wrote using a series of progressive generalizations from the concrete to the abstract (no doubt to appeal to their more experimentally oriented colleagues). See Dominique Pestre, *Physique et physiciens en France, 1918–1940* (Paris, 1984), pp. 18, 56, 117.

37. See Bureau des Longitudes, *Réception des signaux horaires: Renseignements méteorologiques, sismologiques, etc., transmis par les postes de télégraphie sans fil de la tour Eiffel, Lyon, Bordeaux, etc.* (Paris, 1924), pp. 83–84.

38. See Ernest Lebon, *Henri Poincaré: Biographie, bibliographie analytique des écrits* (Paris, 1909), pp. 16–17.

39. Henri Poincaré, "La Mesure du temps," *Revue de métaphysique et de morale* 6 (1898): 11–12, 12, hereafter abbreviated "M"; rpt. in Poincaré, *La Valeur de la science* (1905; Paris, 1970), p. 53; trans. George Bruce Halsted, under the title "The Measure of Time," *The Value of Science* (New York, 1907), p. 35, hereafter abbreviated "MT"; trans. mod.

40. See Bartky, "The Adoption of Standard Time," pp. 25–56.

41. "Poincaré's *Science and Hypothesis* . . . engrossed us and held us spellbound for weeks" (Maurice Solovine, introduction to Einstein, *Letters to Solovine*, trans. Wade Baskin [New York, 1987], p. 9). See, for example, "La Mécanique classique," chap. 6 of Poincaré, *La Science et l'hypothèse* (Paris, 1902), esp. p. 111: "Not only don't we have a direct intuition of the equality of two durations, but we don't even have one of the simultaneity of two different events that occur in two different sites; this is what I explained in an article titled 'The Measure of Time.' "

42. Between 1900 and 1904 Poincaré kept his programmatic statements about simultaneity largely separate from his explorations into the details of electrodynamics. But even when Poincaré did introduce his notion of local time into his electrodynamics to insist on the conventionality of judgments of simultaneity, he did not, as Einstein did, use light-signal coordination to reorganize mechanics and electrodynamics in such a way that force-free analysis of space and time clearly begin before any considerations of electron deformations and molecular forces come into play. For Einstein, it was precisely the point that kinematics, the play of temporal and spatial measures, would enter before dynamics. But here is not the place to sort out the relative contributions of these two physicists. Compare Henri Poincaré, "Relations entre la physique expérimentale et la physique mathématique," in *Rapports présentés au Congrès International de Physique*, ed. Ch.-Éd. Guillaume and L. Poincaré, 4 vols. (Paris, 1900), 1:1–29, and Henri Poincaré, "L'État actuel et l'avenir de la physique mathématique," *Bulletin des sciences mathématiques* 28 (1904): 302–24. For a comparison of Einstein and Poincaré's understanding of the electrodynamics of moving bodies, see Miller, *Albert Einstein's Special Theory of Relativity*, and Pais, *Subtle Is the Lord*.

43. This has been pointed out many times, for instance by Leopold Infeld in *Albert Einstein: His Work and Its Influence on Our Times* (New York, 1950), p. 23; Holton, "Influences on Einstein's Early Work," in *Thematic Origins of Scientific Thought;* and Miller, *Albert Einstein's Special Theory of Relativity* and "The Special Relativity Theory: Einstein's Response to the Physics of 1905," in *Albert Einstein: Historical and Cultural Perspectives,* ed. Holton and Yehuda Elkana (Princeton, N.J., 1982), pp. 3–26.

44. Greg Myers, "From Discovery to Invention: The Writing and Rewriting of Two Patents," *Social Studies of Science* 25 (Feb. 1995): 77.

45. Einstein, "The World as I See It," *Ideas and Opinions,* trans. Sonja Bargmann, ed. Seelig (New York, 1954), p. 10.

46. Favarger, *L'Électricité et ses applications à la chronométrie,* p. 10.

47. Ibid., p. 11.

48. See Galison, *Image and Logic: A Material Culture of Microphysics* (Chicago, 1997).

49. Representations of Einstein's relativity as a culmination of increasingly accurate "no ether" measurements are rife; perhaps the most scholarly attempt to locate Einstein's formulation as a mere variant of the early ether-electron theories is to be found in Edmund Whittaker, *A History of the Theories of Aether and Electricity* (London, 1953), where the chapter "The Relativity Theory of Poincaré and Lorentz" includes the remark: "Einstein published a paper [in 1905] which set forth the relativity theory of Poincaré and Lorentz with some amplifications, and which attracted much attention. He asserted as a fundamental principle the *constancy of the velocity of light* . . . which at the time was widely accepted, but has been severely criticised by later writers" (p. 40). See Holton, *Thematic Origins of Scientific Thought,* esp. chap. 5, and Miller, *Albert Einstein's Special Theory of Relativity.*

50. See Schaffer, "Late Victorian Metrology and Its Instrumentation: A Manufactory of Ohms," in *Invisible Connections: Instruments, Institutions, and Science,* ed. Robert Bud and Susan E. Cozzens (Bellingham, Wash., 1992), pp. 23–59; M. Norton Wise, "Mediating Machines," *Science in Context* 2 (Spring 1988): 77–114; and Galison, *Image and Logic.*

A Designer Universe?

FROM THE NEW YORK REVIEW OF BOOKS

A hypothesis about the formation of the cosmos has sparked a debate that has been taken up by scientists, philosophers and theologians. The "anthropic principle" suggests that the universe was specifically designed to allow for the emergence of intelligent life—amd implies the existence of a designer. Is this claim merely a fancily worded tautology, or is it a useful tool in answering some of the vexatious riddles of the universe's creation? Taking the measure of the anthropic principle, the Nobel laureate physicist Steven Weinberg delivers an eloquent statement in defense of the scientific worldview.

I have been asked to comment on whether the universe shows signs of having been designed. I don't see how it's possible to talk about this without having at least some vague idea of what a designer would be like. Any possible universe could be explained as the work of some sort of designer. Even a universe that is completely chaotic, without any laws or regularities at all, could be supposed to have been designed by an idiot.

The question that seems to me to be worth answering, and perhaps not impossible to answer, is whether the universe shows signs of having been designed by a deity more or less like those of traditional monotheistic religions—not necessarily a figure from the ceiling of the Sistine Chapel, but at least some sort of personality, some intelligence, who created the universe and has some special concern with life, in particular with human life. I suppose that this is not the idea of a designer held by many people today. They may tell me that they are thinking of something much more abstract, some cosmic spirit of order and harmony, as Einstein did. They are certainly free to think that way, but then I don't know why they use words like "designer" or "God," except perhaps as a form of protective coloration.

It used to be obvious that the world was designed by some sort of intelligence. What else could account for fire and rain and lightning and earthquakes? Above all, the wonderful abilities of living things seemed to point to a creator who had a special interest in life. Today we understand most of these things in terms of physical forces acting under impersonal laws. We don't yet know the most fundamental laws, and we can't work out all the consequences of the laws we do know. The human mind remains extraordinarily difficult to understand, but so is the weather. We can't predict whether it will rain one month from today, but we do know the rules that govern the rain, even though we can't always calculate their consequences. I see nothing about the human mind any more than about the weather that stands out as beyond the hope of understanding as a consequence of impersonal laws acting over billions of years.

There do not seem to be any exceptions to this natural order, any miracles. I have the impression that these days most theologians are embarrassed by talk of miracles, but the great monotheistic faiths are founded on miracle stories—the burning bush, the empty tomb, an angel dictating the Koran to Mohammed—and some of these faiths teach that miracles continue at the present day. The evidence for all these miracles seems to me to be considerably weaker than the evidence for cold fusion, and I don't believe in cold fusion. Above all, today we understand that even human beings are the result of natural selection acting over millions of years of breeding and eating.

I'D GUESS that if we were to see the hand of the designer anywhere, it would be in the fundamental principles, the final laws of nature, the book of rules that govern all natural phenomena. We don't know the final laws yet, but as far as we have been able to see, they are utterly impersonal and quite without any special role for life. There is no life force. As Richard Feynman has said, when you look at the universe and understand its laws, "the theory that it is all arranged as a stage for God to watch man's struggle for good and evil seems inadequate."

True, when quantum mechanics was new, some physicists thought that it put humans back into the picture, because the principles of quantum mechanics tell us how to calculate the probabilities of various results that might be found by a human observer. But, starting with the work of Hugh Everett forty years ago, the tendency of physicists who think deeply about these things has been to reformulate quantum mechanics in an entirely objective way, with observers treated just like everything else. I don't know if this program has been completely successful yet, but I think it will be.

I have to admit that, even when physicists will have gone as far as they can go, when we have a final theory, we will not have a completely satisfying picture of the world, because we will still be left with the question "why?" Why this theory, rather than some other theory? For example, why is the world described by quantum mechanics? Quantum mechanics is the one part of our present physics that is likely to survive intact in any future theory, but there is nothing logically inevitable about quantum mechanics; I can imagine a universe governed by Newtonian mechanics instead. So there seems to be an irreducible mystery that science will not eliminate.

But religious theories of design have the same problem. Either you mean something definite by a God, a designer, or you don't. If you don't, then what are we talking about? If you do mean something definite by "God" or "design," if for instance you believe in a God who is jealous, or loving, or intelligent, or whimsical, then you still must confront the question "why?" A religion may assert that the universe is governed by that sort of God, rather than some other sort of God, and it may offer evidence for this belief, but it cannot explain why this should be so.

In this respect, it seems to me that physics is in a better position to give us a partly satisfying explanation of the world than religion can ever be, because although physicists won't be able to explain why the laws of nature are what they are and not something completely different, at least we may be able to explain why they are not *slightly* different. For instance, no one has been able to think of a logically consistent alternative to quantum mechanics that is only slightly different. Once you start trying to make small changes in quantum mechanics, you get into theories with negative probabilities or other logical absurdities. When you combine quantum mechanics with relativity you increase its logical fragility. You find that unless you arrange the theory in just the right way you get nonsense, like effects preceding causes, or infinite probabilities. Religious theories, on the other hand, seem to be infinitely flexible, with nothing to prevent the invention of deities of any conceivable sort.

Now, it doesn't settle the matter for me to say that we cannot see the hand of a designer in what we know about the fundamental principles of science. It might be that, although these principles do not refer explicitly to life, much less human life, they are nevertheless craftily designed to bring it about.

SOME PHYSICISTS HAVE ARGUED that certain constants of nature have values that seem to have been mysteriously fine-tuned to just the values that allow for the possibility of life, in a way that could only be explained by the intervention of a designer with some special concern for life. I am not impressed with these supposed instances of fine-tuning. For instance, one of the most frequently quoted examples of fine-tuning has to do with a property of the nucleus of the carbon atom. The matter left over from the first few minutes of the universe was almost entirely hydrogen and helium, with virtually none of the heavier elements like carbon, nitrogen, and oxygen that seem to be necessary for life. The heavy elements that we find on earth were built up hundreds of millions of years later in a first generation of stars, and then spewed out into the interstellar gas out of which our solar system eventually formed.

The first step in the sequence of nuclear reactions that created the heavy elements in early stars is usually the formation of a carbon nucleus out of three helium nuclei. There is a negligible chance of producing a carbon nucleus in its normal state (the state of lowest energy) in collisions of three helium nuclei, but it would be possible to produce appreciable amounts of carbon in stars if the carbon nucleus could exist in a radioactive state with an energy roughly 7 million electron volts (MeV) above the energy of the normal state, matching the energy of three helium nuclei, but (for reasons I'll come to presently) not more than 7.7 MeV above the normal state.

This radioactive state of a carbon nucleus could be easily formed in stars from three helium nuclei. After that, there would be no problem in producing ordinary carbon; the carbon nucleus in its radioactive state would spontaneously emit light and turn into carbon in its normal nonradioactive state, the state found on earth. The critical point in producing carbon is the existence of a radioactive state that can be produced in collisions of three helium nuclei.

In fact, the carbon nucleus is known experimentally to have just such a radioactive state, with an energy 7.65 MeV above the normal state. At first sight this may seem like a pretty close call; the energy of this radioactive state of carbon misses being too high to allow the formation of carbon (and hence of us) by only 0.05 MeV, which is less than one percent of 7.65 MeV. It may appear that the constants of nature on which the properties of all nuclei depend have been carefully fine-tuned to make life possible.

Looked at more closely, the fine-tuning of the constants of nature here does not seem so fine. We have to consider the reason why the formation of carbon in stars requires the existence of a radioactive state of carbon with an energy not more than 7.7 MeV above the energy of the normal state. The reason is that the carbon nuclei in this state are actually formed in a two-step process: first, two helium nuclei combine to form the unstable nucleus of a beryllium isotope, beryllium 8, which occasionally, before it falls apart, captures another helium nucleus, forming a carbon nucleus in its radioactive state, which then decays into normal carbon. The total energy of the beryllium 8 nucleus and a helium nucleus at rest is 7.4 MeV above the energy of the normal state of the carbon nucleus; so if the energy of the radioactive state of carbon were more than 7.7 MeV it could only be formed in a collision of a helium nucleus and a beryllium 8 nucleus if the energy of motion of these two nuclei were at least 0.3 MeV—an energy which is extremely unlikely at the temperatures found in stars.

Thus the crucial thing that affects the production of carbon in stars is not the 7.65 MeV energy of the radioactive state of carbon above its normal state, but the 0.25 MeV energy of the radioactive state, an unstable composite of a beryllium 8 nucleus and a helium nucleus, above the energy of those nuclei at rest.[1] This energy misses being too high for the production of carbon by a fractional amount of 0.05 MeV/0.25 MeV, or 20 percent, which is not such a close call after all.

THIS CONCLUSION about the lessons to be learned from carbon synthesis is somewhat controversial. In any case, there *is* one constant whose value does seem remarkably well adjusted in our favor. It is the energy density of empty space, also known as the cosmological constant. It could have any value, but from first principles one would guess that this constant should be very large, and could be positive or negative. If large and positive, the cosmological constant would act as a repulsive force that increases with distance, a force that would prevent matter from clumping together in the early universe, the process that was the first step in forming galaxies and stars and planets and people. If large and negative the cosmological constant would act as an attractive force increasing with distance, a force that would almost immediately reverse the expansion of the universe and cause it to recollapse, leaving no time for the evolution of life. In fact, astronomical observations show that the cosmological constant is quite small, very much smaller than would have been guessed from first principles.

It is still too early to tell whether there is some fundamental principle that can explain why the cosmological constant must be this small. But even if there is no such principle, recent developments in cosmology offer the possibility of an explanation of why the measured values of the cosmological constant and other physical constants are favorable for the appearance of intelligent life. According to the "chaotic inflation" theories of André Linde and others, the expanding cloud of billions of galaxies that we call the big bang may be just one fragment of a much larger universe in which big bangs go off all the time, each one with different values for the fundamental constants.

In any such picture, in which the universe contains many parts with different values for what we call the constants of nature, there would be no difficulty in understanding why these constants take values favorable to intelligent life. There would be a vast number of big bangs in which the constants of nature take values unfavorable for life, and many fewer where life is possible. You don't have to invoke a benevolent designer to explain why we are in one of the parts of the universe where life is possible: in all the other parts of the universe there is no one to raise the question.[2] If any theory of this general type turns out to be correct, then to conclude that the constants of nature have been fine-tuned by a benevolent designer would be like saying, "Isn't it wonderful that God put us here on earth, where there's water and air and the surface gravity and temperature are so comfortable, rather than some horrid place, like Mercury or Pluto?" Where else in the solar system other than on earth could we have evolved?

Reasoning like this is called "anthropic." Sometimes it just amounts to an assertion that the laws of nature are what they are so that we can exist, without further explanation. This seems to me to be little more than mystical mumbo jumbo. On the other hand, if there really is a large number of worlds in which some constants take different values, then the anthropic explanation of why in our world they take values favorable for life is just common sense, like explaining why we live on the earth rather than Mercury or Pluto. The actual value of the cosmological constant, recently measured by observations of the motion of distant supernovas, is about what you would expect from this sort of argument: It is just about small enough so that it does not interfere much with the formation of galaxies. But we don't yet know enough about physics to tell whether there are different parts of the universe in which what are usually called the constants of physics really do take different values. This is not a hopeless question; we will be able to answer it when we know more about the quantum theory of gravitation than we do now.

IT WOULD BE EVIDENCE for a benevolent designer if life were better than could be expected on other grounds. To judge this, we should keep in mind that a certain capacity for pleasure would readily have evolved through natural selection, as an incentive to animals who need to eat and breed in order to pass on their genes. It may not be likely that natural selection on any one planet would produce animals who are fortunate enough to have the leisure and the ability to do science and think abstractly, but our sample of what is produced by evolution is very biased, by the fact that it is only in these fortunate cases that there is anyone thinking about cosmic design. Astronomers call this a selection effect.

The universe is very large, and perhaps infinite, so it should be no surprise that, among the enormous number of planets that may support only unintelligent life and the still vaster number that cannot support life at all, there is some tiny fraction on which there are living beings who are capable of thinking about the universe, as we are doing here. A journalist who has been assigned to interview lottery winners may come to feel that some special providence has been at work on their behalf, but he should keep in mind the much larger number of lottery players whom he is not interviewing because they haven't won anything. Thus, to judge whether our lives show evidence for a benevolent designer, we have not only to ask whether life is better than would be expected in any case from what we know about natural selection, but we need also to take into account the bias introduced by the fact that it is we who are thinking about the problem.

This is a question that you all will have to answer for yourselves. Being a physicist is no help with questions like this, so I have to speak from my own experience. My life has been remarkably happy, perhaps in the upper 99.99 percentile of human happiness, but even so, I have seen a mother die painfully of cancer, a father's personality destroyed by Alzheimer's disease, and scores of second and third cousins murdered in the Holocaust. Signs of a benevolent designer are pretty well hidden.

The prevalence of evil and misery has always bothered those who believe in a benevolent and omnipotent God. Sometimes God is excused by pointing to the need for free will. Milton gives God this argument in *Paradise Lost:*

I formed them free, and free they must remain
Till they enthral themselves: I else must change
Their nature, and revoke the high decree
Unchangeable, eternal, which ordained
Their freedom; they themselves ordained their fall.

It seems a bit unfair to my relatives to be murdered in order to provide an opportunity for free will for Germans, but even putting that aside, how does free will account for cancer? Is it an opportunity of free will for tumors?

I don't need to argue here that the evil in the world proves that the universe is not designed, but only that there are no signs of benevolence that might have shown the hand of a designer. But in fact the perception that God cannot be benevolent is very old. Plays by Aeschylus and Euripides make a quite explicit statement that the gods are selfish and cruel, though they expect better behavior from humans. God in the Old Testament tells us to bash the heads of infidels and demands of us that we be willing to sacrifice our children's lives at His orders, and the God of traditional Christianity and Islam damns us for eternity if we do not worship him in the right manner. Is this a nice way to behave? I know, I know, we are not supposed to judge God according to human standards, but you see the problem here: If we are not yet convinced of His existence, and are looking for signs of His benevolence, then what other standards *can* we use?

The issues that I have been asked to address here will seem to many to be terribly old-fashioned. The "argument from design" made by the English theologian William Paley is not on most people's minds these days. The prestige of religion seems today to derive from what people take to be its moral influence, rather than from what they may think has been its success in accounting for what we see in nature. Conversely, I have to admit that, although I really don't believe in a cosmic designer, the reason that I am taking the trouble to argue about it is that I think that on balance the moral influence of religion has been awful.

This is much too big a question to be settled here. On one side, I could point out endless examples of the harm done by religious enthusiasm, through a long history of pogroms, crusades, and jihads. In our own century it was a Muslim zealot who killed Sadat, a Jewish zealot who killed Rabin, and a Hindu zealot who killed Gandhi. No one would say that Hitler was a Christian zealot, but it is hard to imagine Nazism taking the form it did without the foundation provided by centuries of Christian anti-Semitism. On the other side, many admirers of religion would set countless examples of the good done by religion.

For instance, in his recent book *Imagined Worlds,* the distinguished physicist Freeman Dyson has emphasized the role of religious belief in the suppression of slavery. I'd like to comment briefly on this point, not to try to prove anything with one example but just to illustrate what I think about the moral influence of religion.

It is certainly true that the campaign against slavery and the slave trade was greatly strengthened by devout Christians, including the Evangelical layman William Wilberforce in England and the Unitarian minister William Ellery Channing in America. But Christianity, like other great world religions, lived comfortably with slavery for many centuries, and slavery was endorsed in the New Testament. So what was different for anti-slavery Christians like Wilberforce and Channing? There had been no discovery of new sacred scriptures, and neither Wilberforce nor Channing claimed to have received any supernatural revelations. Rather, the eighteenth century had seen a widespread increase in rationality and humanitarianism that led others—for instance, Adam Smith, Jeremy Bentham, and Richard Brinsley Sheridan—also to oppose slavery, on grounds having nothing to do with religion. Lord Mansfield, the author of the decision in *Somersett's Case,* which ended slavery in England (though not its colonies), was no more than conventionally religious, and his decision did not mention religious arguments. Although Wilberforce was the instigator of the campaign against the slave trade in the 1790s, this movement had essential support from many in Parliament like Fox and Pitt, who were not known for their piety. As far as I can tell, the moral tone of religion benefited more from the spirit of the times than the spirit of the times benefited from religion.

Where religion did make a difference, it was more in support of slavery than in opposition to it. Arguments from scripture were used in Parliament to defend the slave trade. Frederick Douglass told in his *Narrative* how his condition as a slave became worse when his master underwent a religious conversion that allowed him to justify slavery as the punishment of the children of Ham. Mark Twain described his mother as a genuinely good person, whose soft heart pitied even Satan, but who had no doubt about the legitimacy of slavery, because in years of living in antebellum Missouri she had never heard any sermon opposing slavery, but only countless sermons preaching that slavery was God's will. With or without religion, good people can behave well and bad people can do evil; but for good people to do evil—that takes religion.

In an e-mail message from the American Association for the Advancement of Science I learned that the aim of this conference is to have a constructive dialogue between science and religion. I am all in favor of a dialogue between science and religion, but not a constructive dialogue. One of the great achievements of science has been, if not to make it impossible for intelligent people to be religious, then at least to make it possible for them not to be religious. We should not retreat from this accomplishment.

NOTES

1. This was pointed out in a 1989 paper by M. Livio, D. Hollowell, A. Weiss, and J. W. Truran ("The anthropic significance of the existence of an excited state of 12C," *Nature,* Vol. 340, No. 6231, July 27, 1989). They did the calculation quoted here of the 7.7 MeV maximum energy of the radioactive state of carbon, above which little carbon is formed in stars.

2. The same conclusion may be reached in a more subtle way when quantum mechanics is applied to the whole universe. Through a reinterpretation of earlier work by Stephen Hawking, Sidney Coleman has shown how quantum mechanical effects can lead to a split of the history of the universe (more precisely, in what is called the wave function of the universe) into a huge number of separate possibilities, each one corresponding to a different set of fundamental constants. See Sidney Coleman, "Black Holes as Red Herrings: Topological fluctuations and the loss of quantum coherence," *Nuclear Physics,* Vol. B307 (1988), p. 867.

About the Contributors

NATALIE ANGIER, whose science writing for the *New York Times* won her the 1991 Pulitzer Prize, started her career as a founding staff member of *Discover* magazine, where she specialized in writing about biology. In 1990, she joined the *Times*, where she has covered genetics, evolutionary biology, medicine, and other subjects. Her work has appeared in a number of major publications, and she is the author of three books: *Natural Obsessions*, about the world of cancer research (recently reissued in a new paperback edition); *The Beauty of the Beastly*; and the national bestseller *Woman: An Intimate Geography*, published originally in 1999 and now available in paperback. She is also the recipient of the American Association for the Advancement of Science-Westinghouse Award for excellence in science journalism and the Lewis Thomas Award for distinguished writing in the life sciences.

"I am no clothes horse," she writes, "and the only time I read *Vogue* magazine is while I'm waiting to get my hair cut, but when the University of Illinois pitched a story to me with the tag line of 'paleofashion,' I couldn't resist. The subject matter was naturally evocative: a novel reinterpretation of the famed 'Venus' figurines, and a heightened appreciation for the role of women's work and women's art in the evolution of prehistoric culture. And Olga Soffer, the anthropologist behind the new revelations about weaving and dress style in the Stone Age, is a writer's dream character, a woman with the fluid sound-biting wit of Dorothy Parker.

"After writing this story, I had the sudden urge to take up needlework. I sat very still—until the crisis passed."

DON ASHER, a writer and jazz pianist, began his professional piano career at age fifteen, in the dives and roadhouses in and around Worcester, Massachusetts. After earning degrees in organic chemistry from Cornell, he worked briefly as a chemist in the mill towns of southern New England, and then returned to his first love, performing in every venue from honky-tonk to society band. From 1960 to 1963, he was house pianist at the storied hungry i nightclub in San Francisco. He is the author of six novels, a collection of stories, two works of nonfiction, and articles for several publications.

Of "Lab Notes," he writes that it "evolved from (1) a desire to explore the circumstances involved in misadventure—people taking wrong turns in their professional lives, and (2) a perverse need to recapture the sights, smells (sweet and awful), and inherent dangers of the chemistry laboratory."

TIMOTHY FERRIS is the author of ten books, among them the bestsellers *Coming of Age in the Milky Way* and *The Whole Shebang*, which have been translated into fifteen languages and were named by the *New York Times* as among the leading books published in the twentieth century. A former newspaper reporter and editor of *Rolling Stone* magazine, he is a frequent contributor to *The New Yorker* and has written and narrated two television specials, *The Creation of the Universe* and *Life Beyond Earth*. He produced the Voyager phonograph record, an artifact of human civilization containing music, sounds of Earth, and encoded photographs launched aboard the Voyager interstellar spacecraft, and was among the journalists selected as candidates to fly aboard the Space Shuttle in 1986. A consultant to NASA on the long-term goals of space exploration, he serves on the space agency's Near-Earth Object Steering Group. The recipient of the American Institute of Physics prize, the American Association for the Advancement of Science Prize, and a Guggenheim Fellowship, he has taught in five disciplines at four universities, and is currently Professor Emeritus at the University of California, Berkeley.

"Since the mid-1970s," he writes, "I've been persuaded that the overwhelming majority of intercourse among intelligent civilizations would be handled not through interstellar spaceflight but via microwave radio or laser communications—and that, if there are many such civilizations in existence at a given time, such communications are likely to be networked. This hypothesis was more difficult to explain back then, before the advent of the Internet and widespread use of global satellite traffic, than it is today. So, when the editors of *Scientific American* asked me to write about interstellar travel, I saw it as an opportunity to show the role that such enterprises might play in a wider context of cosmic communications."

FROM 1983 TO 1992, PETER GALISON taught at Stanford University, where he was in both the philosophy and physics departments. Since 1992, he has been at Harvard University, where he is the Mallinckrodt Professor of the History of Science and of Physics. In 1997, he was named a John D. and Catherine T. MacArthur Foundation Fellow. Galison's main work explores the complex interaction between the three principal subcultures of twentieth-century physics—experimentation, instrumentation, and theory. The volume on experiment (*How Experiments End,* 1987) and that on instrumentation (*Image and Logic,* 1997) are to be followed by the forthcoming final volume, *Theory Machines.* In addition, Galison has launched several projects examining the cross-currents between physics and other fields. These include his coedited volumes on the relations between science, art, and architecture, *The Architecture of Science* (1999) and *Picturing Science, Producing Art* (1998), as well as *Big Science* (1992), *The Disunity of Science* (1996), and *Atmospheric Flight in the 20th Century* (2000).

"For many years," he says, "our picture of Einstein has seemed to me fragmented. On the one side, we know he was a conscientious and quite successful patent examiner. On the other side, and seemingly completely isolated from this first dimension, during those same years Einstein produced three of the pivotal papers in twentieth-century physics. This divided history did not make sense to me, but for many years I saw no way to join them. In the early 1980s (in *How Experiments End*), I finally sorted out some fundamental links between Einstein's patent work and his account of atomic structure—but the relativity story continued to nag at me. Then, just recently, I began thinking about the coordinated clocks that lined the railroad stations, and suddenly an entirely new angle on the problem opened; this essay, 'Einstein's Clocks,' is the first product of those reflections. I am now working this material into a book, tentatively titled *The Empire of Time,* which should be done in about a year."

ATUL GAWANDE is a surgical resident in Boston and a staff writer for *The New Yorker* magazine. He also conducts research on the financing of health care and reduction of error in surgery as a public health researcher at Harvard Medical School. A former laboratory researcher, he has studied retinal disease, the Epstein-Barr virus, and laser surgical techniques. He has also served as a senior health policy adviser in the Clinton administration. He received his M.D. from Harvard Medical School, an M.A. in politics, philosophy, and economics from Oxford University, and an M.P.H. from the Harvard School of Public Health. He lives in Newton, Massachusetts, with his wife, Kathleen Hobson, and three children, Walker, Hattie, and Hunter.

He writes: "Publishing 'When Doctors Make Mistakes' made me very, very nervous. Doctors do not talk about their errors much with one another and not at all in public. There are lawsuits to worry about and a great fear of being regarded as incompetent. But I found there is a science to our failures—and some comfort in understanding them."

DEBORAH M. GORDON is a professor in the department of biological sciences at Stanford University. She attended Oberlin College, Stanford and Duke universities, and did postdoctoral work at Harvard and Oxford before joining the faculty at Stanford. (The website for her laboratory is <http://ant.stanford.edu/welcome.html>) Her essay in this volume is derived from her 1999 book *Ants at Work*.

STEPHEN JAY GOULD is currently Alexander Agassiz Professor of Zoology at Harvard University, where he also holds the positions of professor of geology in the department of earth and planetary science and curator of invertebrate paleontology of the Museum of Comparative Zoology. He is also the Vincent Astor Visiting Research Professor of Biology at New York University and was the 1999–2000 president of the American Association for the Advancement of Science. The recipient of numerous honors and awards, including a John D. and Catherine T. MacArthur Foundation Fellowship, he is well known as the author of the column "This View of Life," which has run monthly in *Natural History* magazine since 1974, and several popular books, among them *The Panda's Thumb*, which won both the American and National Book Awards; *The Mismeasure of Man*, winner of the National Book Critics Circle Award; *Wonderful Life*, which won the Rhone-Poulenc Prize; and, most recently, *Full House* and *Questioning the Millennium*.

"A Division of Worms" originally appeared in two parts in Professor Gould's "This View of Life" column. "This piece is one of my favorites in my essay series," he writes, "because it is based on a real discovery, and then also manages (I think) to express the essence of a very important thinker's views in this context."

STEPHEN S. HALL has been a journalist since the age of sixteen. Currently he is a contributing writer for the *New York Times Magazine*, where he covers biomedicine and the impact of science on the culture at large. His work has appeared in several magazines, and since 1998 he has written a column called "Biology, Inc.," about the biotechnology industry, for *Technology Review*. He is the author of three books on science: *Invisible Frontiers*, about cloning and the birth of the biotech industry; *Mapping the Next Millennium*, a survey of recent scientific work in the fields of geophysics, biology, mathematics, and astronomy within the historical context of mapmaking; and *A Commotion in the Blood*, about the immune system and how it can be enlisted to fight cancer and other diseases.

He writes: "While researching an article on the biology of fear for the *New York Times Magazine*, I underwent a session in a functional MRI machine as part of a scientific study designed to show how the brain's fear center becomes activated when it merely anticipates a fearful experience. I was immediately struck by the power of the technology to provide a glimpse—fleeting but fascinating—of one's own brain at work. I approached Dr. Joy Hirsch of Memorial Sloan-Kettering Cancer Center in New York about the possibilities of performing a series of MRI experiments on my own brain, with an emphasis on exploring cognitive activities involving memory and creativity, including storytelling, thinking up metaphors, and making up sentences. She graciously designed some unique exercises, granted me hours of precious MRI time, and was a true collaborator in every stage of the process. The result was 'Journey to the Center of My Brain,' an exploratory trip into one of science's greatest frontiers, which appeared, appropriately, in a special millennium issue of the *New York Times Magazine* devoted to adventure."

FRANCIS HALZEN is the Hilldale and Gregory Breit Professor in the department of physics at the University of Wisconsin, Madison. A theoretician specializing in particle physics, he is the coauthor, with Alan Martin of the University of Durham, of the textbook *Quarks and Leptons: An Introductory Course in Modern Particle Physics*. He also teaches a laboratory course for art students called "Physics in the Arts."

Regarding his essay in this volume, he explains: "Although by profession a theorist who learned his trade at high-energy particle-accelerator laboratories such as CERN in Geneva and Fermilab near Chicago, I am now spending most of my time working on an experiment in Antarctica. Although particle physics is one of its missions, the Antarctic Muon and Neutrino Detector Array is a telescope that scans the skies for neutrinos. 'Antarctic Dreams' describes the strange, and mostly accidental, circumstances leading to this total metamorphosis of my daily work."

DOUGLAS R. HOFSTADTER is professor of cognitive science at Indiana University, and director of the Center for Research on Concepts and Cognition. He also has connections to several other academic departments, including computer science, psychology, philosophy, and comparative literature. Over the past two decades he has worked toward characterizing the conscious and unconscious cognitive processes underlying both mundane and sophisticated analogies. This has led him to develop a model of what he terms "fluid concepts," which has also amounted to a computational model of the creative process. His acclaimed books include *Gödel, Escher, Bach: an Eternal Golden Braid* (winner of the Pulitzer Prize and The American Book Award), *The Mind's I* (with Daniel Dennett), *Metamagical Themas, Ambigrammi, Fluid Concepts and Creative Analogies, Le Ton beau de Marot,* and a verse translation of Pushkin's *Eugene Onegin.*

He writes that "Analogy as the Core of Cognition" is the "written version of an invited lecture delivered at a major international conference on the topic of analogy and metaphor, which took place in Sofia, Bulgaria, in July 1998. The talk was intended to serve as an antidote to the very narrow view of analogy (which is held, oddly enough, by most members of the cognitive science community) as merely a specialized form of 'reasoning.' " (A fuller version of this same essay can be found in *The Analogical Mind: Perspectives from Cognitive Science,* edited by Dedre Gentner, Keith Holyoak, and Boicho Kokinov, and scheduled for publication by the MIT Press in December 2000.)

BASED IN SANTA FE, NEW MEXICO, GEORGE JOHNSON writes about science for *The New York Times* and in 1999 won the American Association for the Advancement of Science's Science Journalism Award (for large newspapers). He is the author of *In the Palaces of Memory: How We Build the World Inside Our Heads; Fire in the Mind: Science, Faith, and the Search for Order;* and, most recently, *Strange Beauty: Murray Gell-Mann and the Revolution in 20th-Century Physics.* He is codirector of the Santa Fe Science-Writing Workshop and a former Alicia Patterson Fellow. His address on the World Wide Web is <http://talaya.net>.

"A couple of years ago in Santa Fe," he writes, "I listened to a Los Alamos physicist, Geoffrey West, talk about his work on what biologists call the scaling problem. West and two biologists from the University of New Mexico had decided to take on the problem, collaborating in regular meetings at the Santa Fe Institute. The result, as West enthusiastically described it, was a rarity in the world of science: an interdisciplinary project that worked. Their scaling theory, based on the notion of fractal dimensions, uses ideas from the new science of complexity to explain patterns in the biological world. But what interested me as much as the theory was the dynamics of the collaboration. I decided to tell the story as a narrative, a kind of scientific detective story, trying to capture for readers the messy way science is really done."

Susan McCarthy writes about science and the environment; she also writes humor. (When she is very lucky, she is allowed to do both at the same time.) She is a contributing writer to the online magazine *Salon,* and she is the coauthor, with Jeffrey Moussaieff Masson, of *When Elephants Weep: The Emotional Lives of Animals.*

She says that "Must Dog Eat Dog?" "grew from an open-ended assignment on 'trends in sociobiology.' Trying to untangle evolutionary psychology's strands of theory and bias is an endlessly fascinating, endlessly maddening, and probably endless task. Evolutionary biology is like a songwriter who specializes in answer songs, ready to resolve all our questions: Why do fools fall in love? Why don't you treat me like you used to do? How can people be so heartless? How many roads must a man walk down? Where have all the cowboys gone?"

The Onion, founded in 1988, is one of America's most popular humor publications, read by nearly one million people each week. Its 1999 book, *Our Dumb Century,* was a number-one *New York Times* bestseller, and a "best-of" collection, *The Onion's Finest News Reporting* was published by Crown Books in the spring of 2000. *The Onion* is based in Madison, Wisconsin.

Denis G. Pelli studied math as an undergraduate at Harvard and received a Ph.D. in physiology from Cambridge University, where he specialized in vision. From 1981 to 1995 he was a professor at the Institute for Sensory Research at Syracuse University; there, and at the NASA Ames Research Center, he worked on visual requirements of reading and mobility, on visual testing (he helped create the Pelli-Robson Contrast Sensitivity Chart), and on characterizing the limits of visual perception. Since 1995 he has been professor of psychology and neural science at New York University, where he has studied how we see letters and faces as a first step to understanding how we identify objects in general. In August 2000, the Optical Society of America conferred its Leadership Award/New Focus Prize on Dr. Pelli.

"Does size affect shape?" Dr. Pelli asks. After breakthrough work in his lab that showed how letters are seen differently at different sizes, he attended the 1998 Chuck Close retrospective at the Museum of Modern Art. "Looking at the paintings, walking back and forth, seeing the face as flat marks from near and solid from afar, it suddenly dawned on me that Chuck Close's paintings demonstrate that shape depends on size," says Pelli. "And he'd been exhibiting these paintings in public for years. He scooped us! This paper is the result of my efforts over the next year to understand his discovery and help transmit it to the rest of the scientific community. Can art be science? This paper proves one instance of work that qualifies as both."

OLIVER SACKS was born in London in 1933 and educated in London, Oxford, and California. He is a professor of neurology at the Albert Einstein College of Medicine and the author of seven books, including *The Man Who Mistook his Wife for a Hat* and *The Island of the Colorblind*. The feature films *Awakenings* and *At First Sight* were based on Dr. Sacks's work, and he was the host of the BBC *Mind Traveler* series. He lives in New York, where he swims and raises cycads and ferns.

FLOYD SKLOOT'S essays have apeared in *The Best American Essays 1993* and *The Art of the Essay 1999,* and he has recently completed a new collection, *In the Shadow of Memory,* about living with brain damage. His prose and poetry have appeared in *The Atlantic Monthly, Harper's, The American Scholar, Boulevard,* and many other magazines. Story Line Press will publish his second book of poems, *The Evening Light,* in the fall of 2000.

"Gray Area: Thinking with a Damaged Brain" was, he says, "written both to understand and to counteract what has happened to me. By taking on this subject, I knew I was going right to the heart of my weakness, trying to write cogently about thinking with a damaged brain when thinking cogently with a damaged brain is not possible. If I could teach myself to grasp what had occurred, and manage to explain it clearly enough, I felt that in some ways I could defeat it. At the same time I was researching and writing this essay about the way my rational processing capacity had been altered, I was working on a companion essay about the emotional side of the experience—how the emotional life is changed by brain damage. Together, it seemed to me that this work might create a bridge between life before and life after."

———

SHERYL GAY STOLBERG covers medicine and health policy for the Washington bureau of the *New York Times*. She began her newspaper career at the *Journal-Bulletin* in Providence, Rhode Island, and then spent nearly ten years with the *Los Angeles Times* before joining the *New York Times* in 1997. She is a 1983 graduate of the University of Virginia. She lives in Chevy Chase, Maryland, with her husband, photographer Scott Robinson, and their two daughters.

"A few weeks before Jesse Gelsinger died," she writes, "I proposed a story on the broken promise of gene therapy, a field that had been all over the news and then, it seemed to me, disappeared from view. With Jesse's death, the task became obvious: to tell his story against the larger backdrop of the research. I was fortunate to have the cooperation of the Gelsinger family and the scientists, who freely shared details before their work became the subject of intense government scrutiny."

STEVEN WEINBERG is a member of the physics and astronomy departments of the University of Texas, Austin. His research has been honored with numerous prizes and awards, including, in 1979, the Nobel Prize in Physics and, in 1991, the National Medal of Science, as well as election to both the U.S. National Academy of Sciences and Britain's Royal Society, and a dozen honorary doctoral degrees. He is the author of over two hundred scientific articles on elementary particle physics, cosmology, and other subjects, and has also written for such periodicals as *The New York Review of Books*, the *Times Literary Supplement*, *The Atlantic Monthly*, *Time*, and *The New Republic*. His books include *Gravitation and Cosmology: Principles and Applications of the General Theory of Relativity* (1972); *The First Three Minutes* (1977); *Discovery of Subatomic Particles* (1983); *Elementary Particles and The Laws of Physics* (with Richard P. Feynman) (1987); *Dreams of a Final Theory: The Search for the Fundamental Laws of Nature* (1993); and a trilogy, *The Quantum Theory of Fields* (1995, 1996, 2000). His writing on science for the general reader has been honored with the Gemant Award of the American Institute of Physics and the Lewis Thomas Prize for the Scientist as Poet of Rockefeller University. Professor Weinberg was educated at Cornell, Copenhagen, and Princeton, and taught at Columbia, Berkeley, M.I.T., and Harvard before coming to Texas in 1982.

Of "A Designer Universe?" he writes: "This is an edited version of a talk that I was asked to give in April 1999 at the Conference on Cosmic Design of the American Association for the Advancement of Science in Washington, D.C. It was originally published in the October 21, 1999, issue of *The New York Review of Books*."

JONATHAN WEINER has been writing about science and nature since shortly after his graduation from Harvard, in 1976. After working as a senior editor at *The Sciences,* he left in 1985 to write his first book, *Planet Earth,* the companion volume to the PBS series. His 1994 book, *The Beak of the Finch,* won the Pulitzer Prize for General Nonfiction and the *Los Angeles Times* Book Prize for Science. His most recent book, *Time, Love, Memory,* from which the essay in this anthology is excerpted, won the National Book Critics Circle award for general nonfiction. He has taught writing at Princeton University, where he also was a visiting fellow in the department of molecular biology during the writing of *Time, Love, Memory.* He currently is writer-in-residence at Rockefeller University in New York City, and lives in Bucks County, Pennsylvania, with his wife, the children's book author Deborah Heiligman, and their two sons, Aaron and Benjamin.

"Writing my book about Seymour Benzer and his work," he reports, "took me almost five years. By the time I was done I couldn't see how to boil it all down into a profile; that took a couple of months. But I'd always dreamed of writing for *The New Yorker* and when I finally saw the story at a newstand, I felt as if a vanity press must have planted it there. Afterward, reporters and documentary filmmakers called Benzer, but he refused to give any more interviews. He's still going strong in the lab, working on the Methusaleh mutant."